T0192303

Undergraduate Lecture Notes in Physics

Undergraduate Lecture Notes in Physics (ULNP) publishes authoritative texts covering topics throughout pure and applied physics. Each title in the series is suitable as a basis for undergraduate instruction, typically containing practice problems, worked examples, chapter summaries, and suggestions for further reading.

ULNP titles must provide at least one of the following:

- An exceptionally clear and concise treatment of a standard undergraduate subject.
- A solid undergraduate-level introduction to a graduate, advanced, or non-standard subject.
- A novel perspective or an unusual approach to teaching a subject.

ULNP especially encourages new, original, and idiosyncratic approaches to physics teaching at the undergraduate level.

The purpose of ULNP is to provide intriguing, absorbing books that will continue to be the reader's preferred reference throughout their academic career.

More information about this series at http://www.springer.com/series/8917

Giampaolo Cicogna

Exercises and Problems in Mathematical Methods of Physics

Second Edition

Giampaolo Cicogna
Dipartimento di Fisica "Enrico Fermi"
Università di Pisa
Pisa, Italy

ISSN 2192-4791 ISSN 2192-4805 (electronic)
Undergraduate Lecture Notes in Physics
ISBN 978-3-030-59471-8 ISBN 978-3-030-59472-5 (eBook)
https://doi.org/10.1007/978-3-030-59472-5

This Springer imprint is published by the registered company Springer Nature Switzerland AG
The registered company address is: Gewerbestrasse 11, 6330 Cham, Switzerland

Preface to the Second Edition

The main novelty of this second edition concerns the final Chapter of the Answers and Solutions. Solutions which in the first edition were left to the reader have now been added, and—more importantly—a greater amount of details in the answers and additional guidance for arriving at the solutions of the problems has now been inserted, with clear suggestions on the necessary calculations, and precise and stimulating indications on the procedures. In some cases, two or three different ways for solving the problems have been proposed. In addition, the statements of various problems have been carefully rewritten and expanded, and the questions appearing partially incomplete or somewhat unclear in the first edition have been revised and improved.

Finally, the "gray" insertions in the text, providing a useful survey of the principal properties of the various mathematical objects, have been enriched, having always in mind the pedagogical scope of the book, along the lines illustrated in the preface of the first edition.

Pisa, Italy
September 2020

Giampaolo Cicogna

Preface to the First Edition

This book is a collection of 350 exercises and problems in Mathematical Methods of Physics: its peculiarity is that exercises and problems are proposed not in a "random" order, but having in mind a precise didactic scope. Each section and subsection starts with exercises based on first definitions, elementary notions and properties, followed by a group of problems devoted to some intermediate situations, and finally by problems which propose gradually more elaborate developments and require some more refined reasoning.

Part of the problems are unavoidably "routine", but several problems point out nontrivial properties, which are often omitted or only marginally mentioned in the textbooks. There are also some problems in which the reader is guided to obtain some important results which are usually stated in textbooks without complete proofs: for instance, the classical "uncertainty principle" $\Delta t \Delta\omega \geq 1/2$, an introduction to Kramers-Kronig dispersion rules and their relation with causality principles, the symmetry properties of the hydrogen atom and the harmonic oscillator in Quantum Mechanics.

Avoiding unnecessary difficulties and excessive formalism, it offers indeed an alternative way to understand the mathematical notions on which Physics is based, proceeding in a carefully structured sequence of exercises and problems. In this sense, this book may be used as (or perhaps, to some extent, better than) a textbook.

There is no need to emphasize that the best (or perhaps the unique) way to correctly understand Mathematics is that of facing and solving exercises and problems. This holds a fortiori for the present case, where mathematical notions and procedures become fundamental tools for Physics. An example can illustrate perfectly the point. The definition of eigenvectors and eigenvalues of a linear operator needs just two or three lines in a textbook, and the notion is relatively simple and intuitive. But only when one tries to find explicitly eigenvectors and eigenvalues in concrete cases, then one realizes that a lot of different procedures is required and extremely various situations occur. This book offers a fairly exhaustive description of possible cases.

This book covers a wide range of topics useful to Physics: Chap. 1 deals with Hilbert spaces and linear operators. Starting from the crucial concept of complete system of vectors, many exercises are devoted to the fundamental tool provided by Fourier expansions, with several examples and applications, including some typical Dirichlet and Neumann Problems. The second part of the chapter is devoted to studying the different properties of linear operators between Hilbert spaces: their domains, ranges, norms, boundedness, closedness, and to examining special classes of operators: adjoint and self-adjoint operators, projections, isometric and unitary operators, functionals, time-evolution operators. Great attention is paid to the notion of eigenvalues and eigenvectors, with the various procedures and results encountered in their determination. Another frequently raised question concerns the different notions of convergence of given sequences of operators.

Chapter 2 starts with a survey of the basic properties of analytic functions of a complex variable, of their power series expansions (Taylor-Laurent series), of their singularities, including branch points and cut lines. The evaluation of many types of integrals by complex variable methods is proposed. Some examples of conformal mappings are finally studied, in order to solve Dirichlet Problems; the results are compared with those obtained in other chapters with different methods, with a discussion about the uniqueness of the solutions.

The problems in Chap. 3 concern Fourier and Laplace transforms with their different applications. The physical meaning of the Fourier transform as "frequency analysis" is carefully presented. The Fourier transform is extended to the space of tempered distributions $\mathscr{S}'$, which include the Dirac delta, the Cauchy principal part and other related distributions. Applications concern ordinary and partial differential equations (in particular the heat, d'Alembert and Laplace equations, including a discussion about the uniqueness of solutions), and general linear systems. The important notion of Green function is considered in many details, together with the notion of causality. Various examples and applications of Laplace transform are proposed, also in comparison with Fourier transform.

The first problems in Chap. 4 deal with basic properties of groups and group representations. Fundamental results following from Schur lemma are introduced since the beginning in the case of finite groups, with a simple application of character theory, in the study of normal modes and vibrational levels of symmetric systems. Other problems concern the notion and the properties of Lie groups and Lie algebras, mainly oriented to physical examples: rotation groups SO_2, SO_3, SU_2, translations, Euclidean group, Lorentz transformations, dilations, Heisenberg group, SU_3, with their physically relevant representations. The last section starts with some examples and applications of symmetry properties of differential equations, then provides a group-theoretical description of some problems in quantum mechanics: the Zeeman and Stark effects, the Schrödinger equation of the hydrogen atom (the group SO_4), the 3-dimensional harmonic oscillator (the group U_3).

At the end of the book, there are answers and solutions for almost all problems.

This book is the result of my lectures during several decades at the Department of Physics of the University of Pisa. I would like to acknowledge all my colleagues who helped me in the organization of the didactic activity, in the preparation of the problems and for their assistance in the examinations of my students. Special thanks are due to prof. Giovanni Morchio, for his constant invaluable support: many of the problems, specially in Sect. 2 of Chap. 1, have been written with his precious collaboration. I am also grateful to prof. Giuseppe Gaeta for his encouragement to write this book, which follows my previous lecture notes (in Italian) *Metodi Matematici della Fisica*, published by Springer-Verlag Italia in 2008 (second edition in 2015).

Finally, I would thank in advance the readers for their comments, and in particular those readers who will suggest improvements and amendments to all possible misprints, inaccuracies and inadvertent mistakes (hopefully, not too serious) in this book, including also errors and imperfections in my English.

Pisa, Italy
January 2018

Giampaolo Cicogna

Contents

1 Hilbert Spaces ... 1
1.1 Complete Sets, Fourier Expansions ... 1
1.1.1 Preliminary Notions. Subspaces. Complete Sets ... 2
1.1.2 Fourier Expansions ... 9
1.1.3 Harmonic Functions; Dirichlet and Neumann Problems ... 14
1.2 Linear Operators in Hilbert Spaces ... 17
1.2.1 A Survey of General Properties of Operators ... 19
1.2.2 Linear Operators Defined Giving $T e_n = v_n$, and Related Problems ... 21
1.2.3 Operators of the Form $Tx = v(w, x)$ and $Tx = \sum_n v_n(w_n, x)$ ... 29
1.2.4 Operators of the Form $Tf(x) = \varphi(x)f(x)$ ... 34
1.2.5 Problems Involving Differential Operators ... 37
1.2.6 Functionals ... 44
1.2.7 Time-Evolution Problems. Heat Equation ... 46
1.2.8 Miscellaneous Problems ... 51

2 Functions of a Complex Variable ... 57
2.1 Basic Properties of Analytic Functions ... 57
2.2 Evaluation of Integrals by Complex Variable Methods ... 62
2.3 Harmonic Functions and Conformal Mappings ... 70

3 Fourier and Laplace Transforms. Distributions ... 73
3.1 Fourier Transform in $L^1(\mathbf{R})$ and $L^2(\mathbf{R})$ ... 73
3.1.1 Basic Properties and Applications ... 76
3.1.2 Fourier Transform and Linear Operators in $L^2(\mathbf{R})$ ... 82
3.2 Tempered Distributions and Fourier Transforms ... 85
3.2.1 General Properties ... 86
3.2.2 Fourier Transform, Distributions and Linear Operators ... 100

3.2.3 Applications to ODE's and Related Green Functions 103
3.2.4 Applications to General Linear Systems and Green Functions 111
3.2.5 Applications to PDE's 116
3.3 Laplace Transforms 122

4 Groups, Lie Algebras, Symmetries in Physics 129
4.1 Basic Properties of Groups and of Group Representations 129
4.2 Lie Groups and Lie Algebras 134
4.3 The Groups SO_3, SU_2, SU_3 140
4.4 Other Relevant Applications of Symmetries to Physics 142

Answers and Solutions 149

Bibliography 217

Chapter 1
Hilbert Spaces

1.1 Complete Sets, Fourier Expansions

The argument of this section is the study of basic properties of Hilbert spaces, without involving the presence of linear operators.

Among other preliminary mathematical properties, the first problems in Sect. 1.1.1 emphasize the notion of dense subspaces, and the difference between linear subspaces and linear closed *subspaces (i.e., Hilbert subspaces). The fundamental concept of* complete system *(or* complete set*) of vectors is then considered, distinguishing between complete sets and* orthonormal complete sets *(to avoid confusion, the term "basis" is never used in this chapter). A set $\{x_n,\ n = 1, 2, \ldots\}$ (obvious changes if $n \in \mathbf{Z}$) of vectors in a Hilbert space H is complete if the finite linear combinations $\sum_{n=1}^{N} \alpha_n x_n,\ \alpha_n \in \mathbf{C}$, are dense in H; in other words, if for any $x \in H$ and $\varepsilon > 0$ there are an integer N and N complex numbers α_n such that $\|x - \sum_{n=1}^{N} \alpha_n x_n\| < \varepsilon$. A fundamental (and very useful) property of complete sets is that the unique vector in H orthogonal to all vectors x_n is the trivial zero vector. When the complete set is orthonormal, we often change the notation x_n into e_n: then $(e_m, e_n) = \delta_{nm}$; in this case, the "best" choice for the coefficients α_n is $\alpha_n = (e_n, x)$ and any $x \in H$ can be expressed as an "abstract" Fourier expansion:*

$$x = \sum_{n=1}^{\infty} (e_n, x)\, e_n$$

In addition, the Parseval identity $\|x\|^2 = \sum_{n=1}^{\infty} |(e_n, x)|^2$ holds, which generalizes to the infinite-dimensional case the elementary Pythagorean theorem known in $\mathbf{R}^2$ or $\mathbf{R}^3$.

G. Cicogna, *Exercises and Problems in Mathematical Methods of Physics*, Undergraduate Lecture Notes in Physics,
https://doi.org/10.1007/978-3-030-59472-5_1

Many exercises are proposed in the context of "abstract" Hilbert spaces, in the space of numerical sequences ℓ^2, and in the "concrete" space of square-integrable functions L^2.

In the case of spaces of functions $L^2(I)$ where I is an interval of finite length, choosing orthonormal complete sets of trigonometric functions, the above abstract Fourier expansion becomes the classical Fourier expansion, which is, as well-known, a fundamental tool in calculations and applications, see the exercises of Sect. 1.1.2. A special application of Fourier expansion concerns some examples of Dirichlet and Neumann Problems (Sect. 1.1.3).

A remark about the definition of scalar product: in this book, the definition adopted is the one generally used by physicists, i.e., $(\alpha x, y) = \alpha^*(x, y)$ and $(x, \alpha y) = \alpha(x, y)$, $\alpha \in \mathbf{C}$, $x, y \in H$. Similarly, if $H = L^2(I)$, the scalar product is $(f, g) = \int_I f^*(x)g(x)\,dx$. Notice that mathematicians usually adopt the definition $(\alpha x, y) = \alpha(x, y)$ and $(x, \alpha y) = \alpha^*(x, y)$.

1.1.1 Preliminary Notions. Subspaces. Complete Sets

1.1

(1) Consider the sequence of functions defined in $[0, \pi]$

$$f_n(x) = \begin{cases} n \sin nx & \text{for } 0 \le x \le \pi/n \\ 0 & \text{for } \pi/n \le x \le \pi \end{cases} , \qquad n = 1, 2, \dots$$

Show that $f_n(x) \to 0$ *pointwise for all* $x \in [0, \pi]$ (included the point $x = 0$) as $n \to \infty$, but $\int_0^\pi f_n(x)\,dx$ does *not* tend to zero (cf. also Problem 3.22).

(2) Consider a sequence of functions of the form, with $x \in \mathbf{R}$,

$$f_n(x) = \begin{cases} c_n & \text{for } 0 < x < n \\ 0 & \text{elsewhere} \end{cases} , \qquad n = 1, 2, \dots$$

where c_n are constants. Choose c_n in such a way that $f_n(x) \to 0$ uniformly $\forall x \in \mathbf{R}$, but $\int_{-\infty}^{+\infty} f_n(x)\,dx$ does not tend to zero.[1]

(3) Consider sequences of functions of the same form as in (2). Choose (if possible) the constants c_n in such a way that

(*a*) $f_n(x) \to 0$ in the norm $L^2(\mathbf{R})$ but not in the norm $L^1(\mathbf{R})$

(*b*) $f_n(x) \to 0$ in the norm $L^1(\mathbf{R})$ but not in the norm $L^2(\mathbf{R})$

(4) Now consider a sequence of functions of the form, with $x \in \mathbf{R}$,

---

[1] See the Introduction to Sect. 1.2 for the statement of the fundamental Lebesgue theorem about the convergence of the integrals of sequences of functions.

$$f_n(x) = \begin{cases} c_n & \text{for } 0 < x < 1/n \\ 0 & \text{elsewhere} \end{cases}, \qquad n = 1, 2, \ldots$$

where c_n are constants:
(a) verify that $f_n(x) \to 0$ pointwise *almost everywhere*
(b) the same questions as in (3)(a)
(c) the same questions as in (3)(b)

1.2
(1) Show that if a function $f(x) \in L^2(I)$, where I is an interval (of finite length $\mu(I)$), then also $f(x) \in L^1(I)$. Is the converse true? What is the relationship between the norms $\|f\|_{L^1(I)}$ and $\|f\|_{L^2(I)}$?
(2) What changes if $I = \mathbf{R}$?
(3) Is it possible to find a function $f(x) \in L^2(I)$, where e.g. $I = [-1, 1]$, such that $\sup_{x \in I} |f(x)| = \varepsilon$ (where $\varepsilon < 1/2$) but $\|f\|_{L^2(I)} = 1$? Or such that $\|f\|_{L^2(I)} = \varepsilon$ but $\sup_{x \in I} |f(x)| = 1$?
(4) The same questions as in (3) if $I = \mathbf{R}$.
(5) The same questions as in (3) and (4) replacing L^2 with L^1.

1.3
(1) Let $f(x) \in L^2(\mathbf{R})$ and let $f_n(x)$ be the "truncated" functions

$$f_n(x) = \begin{cases} f(x) & \text{for } |x| < n \\ 0 & \text{for } |x| > n \end{cases}, \qquad n = 1, 2, \ldots$$

Show that $f_n \in L^1(\mathbf{R}) \cap L^2(\mathbf{R})$ and $\|f - f_n\|_{L^2} \to 0$ as $n \to \infty$. Conclude: is the subspace of the functions $f \in L^1(\mathbf{R}) \cap L^2(\mathbf{R})$ dense in $L^2(\mathbf{R})$? The same question for the subspace of functions $f \in L^2(\mathbf{R})$ having compact support. Are they Hilbert subspaces in $L^2(\mathbf{R})$?
(2) Is the subspace $\mathscr{S}$ of test functions for the tempered distributions (i.e., the subspace of the C^∞ functions rapidly going to zero with their derivatives as $|x| \to +\infty$) dense in $L^2(\mathbf{R})$?

1.4
(1) Let $g(x) \in L^1(\mathbf{R}) \cap L^2(\mathbf{R})$ be such that

$$\int_{-\infty}^{+\infty} g(x)\, dx = M \neq 0$$

introduce then the functions

$$w_n(x) = \begin{cases} M/n & \text{for } 0 < x < n \\ 0 & \text{elsewhere} \end{cases}, \qquad n = 1, 2, \ldots$$

and let $z_n(x) = g(x) - w_n(x)$. Verify that $\int_{-\infty}^{+\infty} z_n(x)\,dx = 0$ and $\|g - z_n\|_{L^2} \to 0$ as $n \to \infty$.

(2) Conclude: is the set of functions $f(x) \in L^1(\mathbf{R}) \cap L^2(\mathbf{R})$ with zero mean value a *dense* subspace in $L^2(\mathbf{R})$? (for an alternative proof, see Problem 3.8).

1.5

(1) In $L^2(-a, a)$ $(a \neq \infty)$, consider the subspace V of the functions such that

$$\int_{-a}^{a} f(x)\,dx = 0$$

Is this a Hilbert subspace? What is its orthogonal complement, and what are their respective dimensions? Choose an orthonormal complete system in each one of these subspaces. Show that the orthogonal complement $W^{\perp}$ of any subspace W is a closed subspace.

(2) What changes if $a = \infty$? (see previous problem).

(3) The same questions as in (1) and (2) for the subset of the *even* functions such that $\int_{-a}^{a} f(x)\,dx = 0$.

1.6

(1) Consider in the space $L^2(-1, 1)$ the function $u = u(x) = 1$ and consider the sequence of functions

$$g_n(x) = \begin{cases} n|x| & \text{for } \ |x| \le 1/n \\ 1 & \text{for } \ 1 \ge |x| \ge 1/n \end{cases}, \qquad n = 1, 2, \ldots$$

Show that $\|g_n - u\|_{L^2} \to 0$.

(2) Show that the subspace of the functions $g(x) \in L^2(I)$ which are continuous in a neighborhood of a point $x_0 \in I$ and satisfy $g(x_0) = 0$ is dense in $L^2(I)$. Show that the same is also true for the subspace of the functions which are C^{∞} in a neighborhood of $x_0 \in I$ and satisfy $g^{(n)}(x_0) = 0$ for all $n \ge 0$.

1.7

(1) Consider the limit

$$\lim_{N\to\infty} \frac{1}{2N} \int_{-N}^{N} f(x)\,dx$$

Does it exist (and is the same) for all $f(x) \in L^2(\mathbf{R})$?

(2) Consider now the limit

$$\lim_{N\to\infty} \frac{1}{2N} \int_{-N}^{N} x f(x)\,dx$$

Is there a *dense* set of functions $\in L^2(\mathbf{R})$ such that this limit is zero? Find a function $\in L^2(\mathbf{R})$ such that this limit is equal to 1, a function $\in L^2(\mathbf{R})$ such that is $+\infty$; show finally that if $f(x) = \sin(x^{1/3})/x^{2/3}$ this limit does not exist (putting $y = x^{1/3}$ the integral can be easily evaluated).

1.8

(1) Construct a function $f(x) \in L^1(\mathbf{R})$ which *does not* vanish as $|x| \to \infty$. *Hint*: a simple construction is the following: consider a function which is equal to 1 on all intervals $I_n = (n, n + \delta_n)$, $n \in \mathbf{Z}$, $0 < \delta_n < 1$, and equal to zero elsewhere; it is enough to choose suitably $\delta_n \ldots$. With a different choice of δ_n, it is also possible to construct a function $\in L^1(\mathbf{R})$ which is *unbounded* as $|x| \to \infty$.

(2) The same questions for functions $f(x) \in L^2(\mathbf{R})$. It should be clear that the above constructions can be modified in order to have continuous (or even C^∞) functions.

(3) Show that if both $f(x)$ and its derivative $f'(x)$ belong to $L^1(\mathbf{R})$, then $\lim_{|x|\to\infty} f(x) = 0$. *Hint*: it is clearly enough to show that $f(x)$ admits limit at $|x| \to \infty$; to this aim, apply Cauchy criterion: the limit exists if for any $\varepsilon > 0$ one has $|f(x_2) - f(x_1)| < \varepsilon$ for any x_1, x_2 large enough. But

$$f(x_2) - f(x_1) = \int_{x_1}^{x_2} f'(y)\, dy$$

which tends to zero as x_1, x_2 are large enough, because $f'(x) \in L^1(\mathbf{R})$.

(4) Show that if both $f(x)$ and its derivative $f'(x)$ belong to $L^2(\mathbf{R})$, then $\lim_{|x|\to\infty} f(x) = 0$. *Hint*: use the same criterion (assume for simplicity $f(x)$ real):

$$f^2(x_2) - f^2(x_1) = \int_{x_1}^{x_2} \frac{d}{dy} f^2(y)\, dy = \cdots$$

1.9

(1) Show that any sequence of *orthonormal* (not necessarily complete) elements $\{e_n, n = 1, 2, \ldots\}$, in a Hilbert space H is not norm-convergent (check the Cauchy property) as $n \to \infty$, but weakly convergent (to what vector?).

(2) Let $\{x_n, n = 1, 2, \ldots\}$ be any sequence of vectors:
(a) show that if there is some $x \in H$ such that $\|x_n\| \to \|x\|$ and x_n weakly converges to x, then x_n is norm convergent to x, i.e. $\|x_n - x\| \to 0$.
(b) show that if x_n is norm convergent to x, then the sequence x_n is bounded, i.e., there is a positive constant M such that $\|x_n\| < M, \ \forall n$.

1.10

Consider the following linear subspaces of the Hilbert space $L^2(-1, 1)$:

$$V_1 = \left\{ \text{the even polynomials, i.e. the polynomials of the form} \sum_{n=0}^{N} a_n x^{2n} \right\} ;$$

$$V_2 = \{\text{the even } C^\infty \text{ functions}\} ;$$

$$V_3 = \left\{ \text{the functions } g(x) \text{ such that } \int_0^1 g(x)\, dx = 0 \right\} ;$$

$$V_4 = \{\text{the functions } g(x) \in C^0 \text{ such that } g(0) = 0\} .$$

What of these subspaces is a *Hilbert* subspace? and what are their respective orthogonal complementary subspaces?

1.11

Recalling that $\{x^n,\ n = 0, 1, 2, \ldots\}$ is a complete set in $L^2(-1, 1)$:

(1) Deduce: is the set of polynomials a dense subspace in $L^2(-1, 1)$? Is a Hilbert subspace?

(2) Show that $\{x^{2n}\}$ is a complete set in $L^2(0, 1)$. And $\{x^{2n+1}\}$?

(3) Show that also $\{x^N,\ x^{N+1},\ x^{N+2}, \ldots\}$, where N is *any* fixed integer > 0, is a complete set in $L^2(-1, 1)$.

1.12

Let $\{e_n,\ n = 1, 2, \ldots\}$ be an orthonormal complete system in a Hilbert space H.

(1) Is the set $v_n = e_n - e_1,\ n = 2, 3, \ldots$ a complete set in H?

(2) Fixed any $w \in H$, is the set $v_n = e_n - w$ a complete set in H?

(3) Let w be any nonzero vector: for what sequences of complex numbers α_n is the set $v_n = e_n - \alpha_n w$ *not* a complete set in H?

(4) Under what condition on $\alpha,\ \beta \in \mathbf{C}$ (with $\alpha\,\beta \neq 0$) is the set $v_n = \alpha e_n - \beta e_{n+1}$ a complete set in H?

1.13

Let $\{e_n,\ n \in \mathbf{Z}\}$ be an orthonormal complete system in a Hilbert space H.

(1) Is the set $v_n = e_n - e_{n+1}$ a complete set in H? And the set $w_n = \alpha e_n - \beta e_{n+1}$ where $\alpha,\ \beta \in \mathbf{C}$?

(2) Let now $H = L^2(0, 2\pi)$ and $e_n = \exp(inx)/\sqrt{2\pi}$: the sets $v_n,\ w_n$ acquire a "concrete" form. Confirm the results obtained above.

1.14

(1) Specify what among the following sets, with $n = 1, 2, \ldots$, are complete in $L^2(-\pi, \pi)$:

$$(a)\ \{x,\ x \cos nx,\ x \sin nx\} ;\ (b)\ \{P(x),\ P(x) \cos nx,\ P(x) \sin nx\}$$

where $P(x)$ is a polynomial (does the answer depend on the form of $P(x)$?);

$$(c)\ \{1,\ x\cos nx,\ \sin nx\}\ ;\ (d)\ \{x,\ \cos nx,\ \sin nx\}\ ;\ (e)\ \{x^2,\ \cos nx,\ \sin nx\}\ ;$$

$$(f)\ \{x,\ x\cos nx,\ \sin n|x|\}\ ;\ (g)\ \{x\cos nx,\ x\sin nx\}\ ;\ (h)\ \{x^{1/3}\cos nx,\ x^{1/3}\sin nx\}$$

(2) If $\{e_n(x)\}$ is a complete set in $H = L^2(I)$, under what conditions for the function $h(x)$ is the set $\{h(x)e_n(x)\}$ complete in H?

1.15

(1) Let $\{a_n,\ n = 1, 2, \ldots\}$ be a sequence of complex numbers $\in \ell^1$, i.e., such that $\sum_{n=1}^{\infty} |a_n| < \infty$. Show that also $\{a_n\} \in \ell^2$, i.e., $\sum_{n=1}^{\infty} |a_n|^2 < \infty$. Is the converse true?

(2) Show that the space ℓ^1 is a dense subspace in the Hilbert space ℓ^2.

1.16

In the space ℓ^2 consider the set, with $n = 1, 2, \ldots,$

$$w_1 = (1, -1, 0, 0, \ldots)/\sqrt{2}\ ,\quad w_2 = (1, 1, -2, 0, 0, \ldots)/\sqrt{6}\ ,\quad \ldots,$$

$$w_n = (\underbrace{1, 1, \ldots, 1}_{n}, -n, 0, 0, \ldots)/\sqrt{n(n+1)}\ ,\quad \ldots$$

(1) Show that this is an orthonormal complete system in ℓ^2.

(2) Deduce that the subspace $\ell^{(0)} \subset \ell^2$ of the sequences such that $\sum_{n=1}^{\infty} a_n = 0$ is *dense* in ℓ^2.

(3) Show that

$$z_n = (1, \underbrace{-1/n, \ldots, -1/n}_{n}, 0, 0, \ldots) \in \ell^{(0)}$$

and that $z_n \to e_1 = (1, 0, 0, \ldots)$ as $n \to \infty$.

1.17

(1) In the space $H = L^2(0, +\infty)$ consider the set of orthonormal functions

$$u_n(x) = \begin{cases} 1 & \text{for}\ \ n-1 < x < n \\ 0 & \text{elsewhere} \end{cases}\ ,\qquad n = 1, 2, \ldots$$

Here are three possible answers to the question: Is this set a complete set in H? What is the correct answer?

(α) the condition $(u_n, f) = 0,\ \forall n$ is $\int_{n-1}^{n} f(x)\,dx = 0,\ \forall n$, and this happens only if $f = 0$, then the set is complete.

(β) the function $f(x) = \sin 2\pi x$ if $x \geq 0$ satisfies $(u_n, f) = 0,\ \forall n$, then the set is not complete.

(γ) the function (e.g.) $f(x) = \begin{cases} \sin 2\pi x & \text{for } 0 < x < 1 \\ 0 & \text{for } x > 1 \end{cases}$ satisfies $(u_n, f) = 0,\ \forall n$,

then the set is not complete.

(2) In the same space consider the set of orthogonal functions

$$v_n(x) = \begin{cases} \sin x & \text{for } (n-1)\pi \leq x \leq n\pi \\ 0 & \text{elsewhere} \end{cases} \quad , \qquad n = 1, 2, \ldots$$

Is this set complete in H?

(3) In the same space consider the set of functions

$$w_n(x) = \begin{cases} \sin nx & \text{for } 0 \leq x \leq n\pi \\ 0 & \text{for } \ x \geq n\pi \end{cases} \quad , \qquad n = 1, 2, \ldots$$

(*a*) Are the functions $w_n(x)$ orthogonal?
(*b*) Is the set $\{w_n(x)\}$ a complete set?

1.18
(1) Is the set $\{x \sin nx,\ n = 1, 2, \ldots\}$ a complete set in $L^2(0, \pi)$? And the subset $\{x \sin nx\}$ with $n = 2, 3, \ldots$?

(2) The same questions for the set $\{x^2 \sin nx,\ n = 1, 2, \ldots\}$ and resp. for the subset $\{x^2 \sin nx\}$ with $n = 2, 3, \ldots$

1.19
(*a*) Is the set $\{\exp(-nx)\,, n = 1, 2, \ldots\}$ a complete set in $L^2(0, +\infty)$? *Hint*: put $y = \exp(-x)$.

(*b*) The same for the set $\{\exp(inx),\ n \in \mathbf{Z}\}$ in $L^2(-2\pi, 2\pi)$

(*c*) The same for the set $\{\exp(inx),\ n \in \mathbf{Z}\}$ in $L^2(0, \pi)$

(*d*) The same for the set $\{\sin nx\ \sin ny\}$, $n = 1, 2, \ldots$ in $L^2(Q)$, where Q is the square $0 \leq x \leq \pi,\ 0 \leq y \leq \pi$

(*e*) The same for the set $\{\exp(-nx)\ \sin ny\}$, $n = 1, 2, \ldots$ in $L^2(\Omega)$, where Ω is the semi-infinite strip $x \geq 0,\ 0 \leq y \leq \pi$

(*f*) The same for the set $\{\exp(-x^2) \exp(inx),\ n \in \mathbf{Z}\}$ in $L^2(\mathbf{R})$.

(Other examples of complete sets in $L^2(0, +\infty)$ will be proposed in Problem 3.131: the proofs are based on properties of Laplace transform and complex functions.)

1.1.2 Fourier Expansions

1.20

(1) Evaluate the Fourier expansion in terms of the orthonormal complete system in $L^2(-\pi,\pi)$

$$\frac{1}{\sqrt{2\pi}},\ \frac{1}{\sqrt{\pi}}\sin nx,\ \frac{1}{\sqrt{\pi}}\cos nx\,, \qquad n=1,2,\ldots$$

of the following functions:

$$f_1(x)=\begin{cases}-1 \text{ for } -\pi<x<0\\ 1 \text{ for } \quad 0<x<\pi\end{cases} \quad ; \qquad f_2(x)=|x|$$

and discuss the convergence of the series.

(2)(*a*) Put $x=\pi/2$ in the expansion of $f_1(x)$ and deduce the sum of the series $1-\frac{1}{3}+\frac{1}{5}-\cdots$

(*b*) Put $x=\pi$ in the expansion of $f_2(x)$ and deduce the sum of the series $1+\frac{1}{9}+\frac{1}{25}+\cdots$

(3) The same questions in (1) for the function in $L^2(0,\pi)$

$$f(x)=1$$

in terms of the orthonormal complete system

$$\sqrt{\frac{2}{\pi}}\sin nx\,, \qquad n=1,2,\ldots$$

Notice that the series is automatically defined $\forall x\in\mathbf{R}$, also out of the interval $0,\pi$: to what function does this series converge? Does it converge at the point $x=\pi$? to what value? and at the point $x=3\pi/2$? to what value?

1.21

(1) Find the Fourier expansion in terms of the complete set $\{1,\ \cos nx,\ \sin nx\,, n=1,2,\ldots\}$ in $L^2(-\pi,\pi)$ of the function $f_1(x)=x$ (with $-\pi<x<\pi$) and discuss the convergence of the series.

(2) The same for the function

$$f_2(x)=\begin{cases}x+\pi & \text{for} \quad -\pi<x<0\\ x-\pi & \text{for} \quad\ \ 0<x<\pi\end{cases}$$

Recognize that the two functions $f_1(x)$ and $f_2(x)$ (or better, their periodic prolongations with period 2π) are actually the same function apart from a translation; accordingly, verify that their Fourier coefficients are related by a simple rule.

1.22

In the space $L^2(0,a)$ the following three sets are, as well-known, orthogonal complete sets

$$(i) \quad 1,\ \cos\left(\frac{2n\pi}{a}x\right),\ \sin\left(\frac{2n\pi}{a}x\right), \qquad n = 1, 2, \dots$$

$$(ii)\ 1,\ \cos\left(\frac{n\pi}{a}x\right) \ ; \ (iii)\ \sin\left(\frac{n\pi}{a}x\right), \qquad n = 1, 2, \dots$$

Verify that the series obtained as Fourier expansion of a function $f(x) \in L^2(0,a)$ with respect to the set (i) is automatically extended to all $x \in \mathbf{R}$ and converges to a function $\widetilde{f_1}(x)$ with period a, whereas the series obtained as Fourier expansion with respect to the sets (ii) and (iii) converge to functions $\widetilde{f_2}(x)$ and $\widetilde{f_3}(x)$ with period $2a$. Consider for instance the function $f(x) = x \in L^2(0,a)$: without evaluating its Fourier expansions, specify what are the functions $\widetilde{f_1}(x)$, $\widetilde{f_2}(x)$, $\widetilde{f_3}(x)$.

1.23

Consider the space $L^2(Q)$, where Q is the square $0 \le x \le \pi,\ 0 \le y \le \pi$.

(1) Evaluate the double Fourier expansion of the function

$$f(x,y) = 1$$

in terms of the orthonormal complete system

$$e_{n,m} = \frac{2}{\pi} \sin nx \sin my\,, \qquad n, m = 1, 2, \dots$$

The series is automatically defined $\forall x, y \in \mathbf{R}^2$: to what function $\widetilde{f}(x,y)$ does this series converge?

(2) The same questions for the function

$$f(x,y) = \sin x$$

1.24

(1) Show that if the coefficients a_n of a Fourier series in $L^2(0,2\pi)$ of the form

$$f(x) = \sum_{n=-\infty}^{+\infty} a_n \exp(inx)$$

satisfy $\sum_n |a_n| < \infty$, i.e. $\{a_n\} \in \ell^1$, then $f(x)$, with its periodic prolongation out of the interval $(0, 2\pi)$, is continuous.

(2) Generalize: assume that for some integer h one has

$$\sum_{n=-\infty}^{+\infty} |n^h a_n| < \infty$$

How many times (at least) is the function $f(x)$ continuously differentiable?

(3) Assume that for some real α one has

$$|a_n| \le \frac{c}{|n|^\alpha} \quad \text{with} \quad \alpha > k + \frac{1}{2}$$

at least for $|n| > n_0$ where n_0 and k are given integers and c a constant. Show that $f(x) \in C^{k-1}$, i.e. $f(x)$ is $k-1$ times continuously differentiable, and that $f^{(k)}(x)$ is possibly not continuous but $\in L^2(0, 2\pi)$.

(4) Assume that the coefficients a_n satisfy a condition of the form, if $|n| > n_0$,

$$|a_n| \le \frac{c}{2^{|n|}}$$

what property of differentiability can be expected for the function $f(x)$? (Clearly, all the above results also hold for similar Fourier expansions where $\exp(inx)$ are replaced e.g. by $\cos nx$ and/or $\sin nx$.)

1.25

(1) Specify what properties can be deduced for the function $f(x) \in L^2(-\pi, \pi)$ if its Fourier series is

$$f(x) = \sum_{n=1}^{\infty} \frac{1}{n^2(n^3+1)^{1/4}} \cos nx$$

(2) Show that any function $f(x)$ admitting a Fourier series of the following form

$$f(x) = \sum_{n=1}^{\infty} \frac{a_n}{n} \sin nx$$

where $a_n \in \ell^2$, is a continuous function, included its periodic prolongation out of the interval $(-\pi, \pi)$ (extensions to series of similar form, where $\sin nx$ is replaced, e.g., by $\cos nx$ or $\exp(inx)$ are obvious).

1.26

In all the questions of this problem, do *not* try to evaluate the Fourier coefficients of the proposed functions. No calculations needed!

(1) In $H = L^2(-\pi, \pi)$ consider the function

$$f(x) = \begin{cases} 0 & \text{for } -\pi \le x \le 0 \\ x\sqrt{x} & \text{for } \ 0 \le x \le \pi \end{cases}$$

(a) Is the Fourier expansion of $f(x)$ with respect to the complete set $\{\exp(inx),\ n \in \mathbf{Z}\}$ convergent at the point $x = \pi$? To what value? and at the point $x = 4$?
(b) Is it true that the Fourier coefficients c_n of the above expansion satisfy $\{c_n\} \in \ell^1$?

(2) In the same space H let $f(x) = \sqrt{|x|}$. Is it true that the Fourier coefficients a_n of the expansion $f(x) = \sum_n a_n \cos nx$ satisfy $\{na_n\} \in \ell^2$?

(3) In the same space H let $f(x) = \exp(x^2)$. Is it true that the Fourier coefficients a_n of the expansion $f(x) = \sum_n a_n \cos nx$ satisfy $\{na_n\} \in \ell^2$? and $\{na_n\} \in \ell^1$?

1.27
Let v_n be the orthonormal complete system in $L^2(0, \pi)$

$$v_n(x) = \sqrt{2/\pi}\,\sin nx\,, \qquad n = 1, 2, \ldots$$

(1) Consider the Fourier expansion of the function

$$f_1(x) = \left|x - (\pi/2)\right|$$

with respect to the subset $v_1, v_3, \ldots, v_{2m+1}, \ldots$: is this expansion convergent (with respect to the L^2 norm, of course)? to what function? (No calculation needed!)

(2) The same questions for the function $f_2(x) = x - (\pi/2)$.

(3) The same questions for the function $f_3(x) = x$.

(4) Is the subset $v_1, v_3, \ldots, v_{2m+1}, \ldots$ a complete system in the space $L^2(0, \pi/2)$?

1.28
Consider in the space $L^2(-\pi, \pi)$ the orthonormal *not* complete set

$$\frac{1}{\sqrt{2\pi}},\ \frac{1}{\sqrt{\pi}}\sin nx\,, \qquad n = 1, 2, \ldots$$

Find the functions which are obtained performing the Fourier expansion (no calculation needed!) of the following functions with respect to this set:

$$f_1(x) = 2 + \exp(2ix)\,,\quad f_2(x) = x\log|x|\,,\quad f_3(x) = \begin{cases} 1 & \text{for } |x| < \pi/2 \\ 0 & \text{for } \pi/2 < |x| < \pi \end{cases}$$

1.29
Consider in the space $L^2(0, 4\pi)$ the orthogonal *not* complete set $\{\exp(inx), n \in \mathbf{Z}\}$. Find the functions which are obtained performing the Fourier expansion (no calculation needed!) of the following functions with respect to this set:

$$f_1(x) = \begin{cases} 1 & \text{for } 0 < x < 2\pi \\ 0 & \text{for } 2\pi < x < 4\pi \end{cases} \quad ; \quad f_2(x) = |\sin(x/2)|$$

1.30
Consider in $L^2(0, \infty)$ the set

$$v_n(x) = \begin{cases} \sin nx & \text{for } 0 \le x \le n\pi \\ 0 & \text{for } x \ge n\pi \end{cases} , \qquad n = 1, 2, \ldots$$

See Problem 1.17, q.(3) for the orthogonality and the non-completeness of this set. Find the functions which are obtained performing the Fourier expansion of the following functions with respect to this set:

$$f_1(x) = \begin{cases} 1 & \text{for } \ 0 < x < \pi \\ 0 & \text{for } \ x > \pi \end{cases} \quad ; \quad f_2(x) = \begin{cases} 1 & \text{for } \ 0 < x < 2\pi \\ 0 & \text{for } \ x > 2\pi \end{cases}$$

1.31
In the space $H = L^2(0, +\infty)$ consider the set of orthonormal functions $u_n(x)$

$$u_n(x) = \begin{cases} 1 & \text{for } \ n-1 < x < n \\ 0 & \text{elsewhere} \end{cases} , \qquad n = 1, 2, \ldots$$

already proposed in Problem 1.17, q.(1).

(1) What function is obtained performing the Fourier expansion of a function $f(x) \in L^2(0, \infty)$ with respect to the set $u_n(x)$?
(2) Is the sequence of the functions $u_n(x)$ pointwise convergent as $n \to \infty$? Is the convergence uniform? Is this sequence a Cauchy sequence (with respect to the $L^2(0, \infty)$ norm)? Is it *weakly* L^2-convergent (i.e., does the numerical sequence (u_n, g) admit limit $\forall g \in L^2(0, \infty)$)?

In the 3 following problems, the independent variable is the time t —just to help the physical interpretation—instead of the "position" variable x. Accordingly, we write, e.g., $u = u(t)$, $\dot{u} = du/dt$, etc.

1.32
(1) Consider the equation of the periodically forced harmonic oscillator

$$\ddot{u} + u = g(t)\,, \qquad u = u(t)$$

where $g(t)$ is a 2π-periodic given function $\in L^2(0, 2\pi)$, and look for 2π-periodic solutions $u(t)$. Write $g(t)$ as Fourier series with respect to the orthogonal complete system $\{\exp(int),\ n \in \mathbf{Z}\}$ in $L^2(0, 2\pi)$: $g(t) = \sum_n g_n \exp(int)$, and obtain the solution in the form of a Fourier series: $u(t) = \sum_n u_n \exp(int)$. Under what condition on

$g(t)$ (or on its Fourier coefficients g_n) does this equation admit solution? and, when the solution exists, is it unique?

(2) The same questions for the equation

$$\ddot{u} + 2u = g(t)$$

1.33

(1) The same questions as in q. (1) of the above problem for the equation

$$\dot{u} + u = g(t), \qquad u = u(t)$$

(this is, e.g., the equation of an electric series circuit of a resistance R and an inductance L (with $R = L = 1$), submitted to a periodic potential $g(t)$, where $u(t)$ is the electric current). As before, assume that $g(t)$ is a 2π-periodic given function $\in L^2(0, 2\pi))$ and look for 2π-periodic solutions $u(t)$. Introducing the orthogonal complete system $\{\exp(int),\ n \in \mathbf{Z}\}$ in $L^2(0, 2\pi)$, write in the form of a Fourier series the solution of this equation.

(2) Show that the solution $u(t)$ is a continuous function.

1.34

A doubt concerning the existence and uniqueness of solutions of the equations given in the two above problems. In Problem 1.32, q.(1) the conclusion was that the equation $\ddot{u} + u = g$ has no solution if the Fourier coefficients $g_{\pm 1}$ of $g(t)$ with respect the orthogonal complete set $\{\exp(int),\ n \in \mathbf{Z}\}$ are not zero. However, it is well-known from elementary analysis that e.g. the equation $\ddot{u} + u = \sin t$ admits the solution $u(t) = -(t/2)\cos t$ (this is the case of "resonance"). Explain why this solution does not appear in the present context of Fourier expansions. A related difficulty appears in Problems 1.32, q.(2) and 1.33, q.(1): the conclusion was that the solution is unique, but it is well-known that the differential equations considered in these problems admit resp. ∞^2 and ∞^1 solutions: explain why these solutions do not appear in the above calculations. Similar apparent difficulties appear in many other cases: see e.g. Problems from 1.91–1.97.

1.1.3 *Harmonic Functions; Dirichlet and Neumann Problems*

In this subsection simple examples of Dirichlet and Neumann problems will be proposed. The Dirichlet problem amounts of finding a harmonic function $U = U(x, y)$ in some region $\Omega \subset R^2$ satisfying a given condition on the boundary *of Ω, i.e.:*

$$\Delta_2 U \equiv \frac{\partial^2 U}{\partial x^2} + \frac{\partial^2 U}{\partial y^2} = 0 \text{ in } \Omega, \text{ with } U\Big|_{\partial\Omega} = F(x, y)$$

Neumann Problem amounts of finding a harmonic function in Ω when a condition is given on the normal derivative on the boundary: $\partial U/\partial n\big|_{\partial\Omega} = G(x, y)$.

The Dirichlet Problem will be also reconsidered, with different methods, in Chap. 2, Sect. 2.3, and in Chap. 3, Problems 3.110–3.112, 3.115.

In the 4 following exercises, recall that the most general form of a harmonic function $U = U(r, \varphi)$ in the interior of the circle centered at the origin of radius R, in polar coordinates r, φ, is given by

$$U(r, \varphi) = a_0 + \sum_{n=1}^{\infty} r^n (a_n \cos n\varphi + b_n \sin n\varphi)$$

1.35

(1) Assume for simplicity $R = 1$. Solve the Dirichlet Problem for the circle, i.e., find $U(r, \varphi)$ for $r < 1$ if the boundary value $U(1, \varphi) = F(\varphi) \in L^2(0, 2\pi)$ is given:
(a) if $F(\varphi) = 1$ (trivial!)
(b) if $F(\varphi) = \cos^2 \varphi$ (nearly trivial!)
(c) obtain $U(r, \varphi)$ as a Fourier series if $F(r, \varphi) = \begin{cases} 1 & \text{for } 0 < \varphi < \pi \\ -1 & \text{for } \pi < \varphi < 2\pi \end{cases}$.

(2) Show that $U(r, \varphi)$ is a C^∞ function if $r < 1$.

1.36

Consider the case of a semicircle $0 \le \varphi \le \pi$ (radius $R = 1$) with the boundary conditions

$$U(r, 0) = U(r, \pi) = 0, \quad U(1, \varphi) = F(\varphi) \in L^2(0, \pi)$$

(1) Show that the Dirichlet Problem can be solved with $a_0 = a_n = 0$ for all n.

(2) Let $F(\varphi) = 1$: the solution $U(r, \varphi)$ (written as a Fourier series) can be also extended to the semicircle with $\pi < \varphi < 2\pi$. What is the value of $U(1, 3\pi/2)$?

(3) Solve the Dirichlet Problem with the boundary conditions

$$U(r, 0) = U(r, \pi) = a \neq 0, \quad U(1, \varphi) = F(\varphi) \in L^2(0, \pi)$$

where $a =$ const. *Hint*: solve first the problem with $U(1, \varphi) = F(\varphi) - a$ and $U(r, 0) = U(r, \pi) = 0$, then ....

1.37

Consider the case of a quarter-circle $0 \le \varphi \le \pi/2$ (radius $R = 1$) with the boundary conditions

$$U(r, 0) = U(r, \pi/2) = 0, \quad U(1, \varphi) = F(\varphi) \in L^2(0, \pi/2)$$

(1) Show that in this case the Dirichlet Problem can be solved with $a_0 = a_n = 0$ for all n, and $b_n = 0$ if n is odd.

(2) Let $F(\varphi) = 1$: the solution $U(r, \varphi)$ (written as a Fourier series) can be also extended to the whole circle. What is the value of $U(1, 3\pi/4)$? and $U(1, 5\pi/4)$? and $U(1, 7\pi/4)$?

1.38

(1) Show that the Neumann Problem for the circle, i.e., the problem of finding $U(r, \varphi)$ in the interior of the circle if the normal derivative at the boundary $\partial U/\partial r|_{r=R} = G(\varphi) \in L^2(0, 2\pi)$ is given, can be solved if (and only if) $G(\varphi)$ satisfies

$$g_0 = \frac{1}{2\pi} \int_0^{2\pi} G(\varphi)\, d\varphi = 0$$

Show also that the solution (when it exists) is not unique.

(2) If $U(r, \varphi)$ is a 2-dimensional electric potential, explain why the results obtained in (1) admit a clear physical interpretation.

1.39

(1) Consider the case of a rectangle in the (x, y) plane, say $0 \le x \le \pi$, $0 \le y \le h$, with boundary conditions

$$U(0, y) = U(\pi, y) = 0, \quad U(x, 0) = F_1(x), \quad U(x, h) = F_2(x)$$

Using the separation of variables $U(x, y) = X(x)Y(y)$, show that the solution of the Dirichlet Problem can be written in the form

$$U(x, y) = \sum_{n=1}^{\infty} \sin nx \big(a_n \exp(ny) + b_n \exp(-ny)\big)$$

where the coefficients a_n, b_n are uniquely determined by $F_1(x)$, $F_2(x)$.

(2) Find $U(x, y)$ in the case $h = 1$ and $F_1(x) = \sin x$, $F_2(x) = \sin 2x$.

(3) What changes if $h = \infty$ (imposing that the solution belongs to L^2)?

1.40

Consider the Dirichlet Problem in a rectangle with nonzero boundary conditions on all the four sides of the rectangle. Show how the problem can be solved by a superposition of two problems similar to the previous one, q. (1).

1.41

(1) Consider the Dirichlet Problem in the annular region between the two circles centered at the origin with radius $R_1 < R_2$. Recalling that the most general form of the harmonic function in the region $R_1 < r < R_2$ can be written in polar coordinates r, φ as

$$U(r,\varphi) = a_0 + b_0 \log r + \sum_{n=\pm1,\pm2,\ldots} \exp(in\varphi)(a_n r^n + b_n r^{-n})$$

show that the Dirichlet Problem admits unique solution imposing the two boundary conditions

$$U(R_1,\varphi) = F_1(\varphi), \quad U(R_2,\varphi) = F_2(\varphi)$$

(2) Solve the problem in the (rather simple) cases
(a) $F_1(\varphi) = c_1,\ F_2(\varphi) = c_2 \neq c_1$, where $c_1,\ c_2$ are constants
(b) $F_1(\varphi) = \cos\varphi$ with $R_1 = 1/2$ and $F_2(\varphi) = \cos\varphi$ with $R_2 = 2$
(c) $F_1(\varphi) = \cos\varphi$ with $R_1 = 1$ and $F_2(\varphi) = \cos 2\varphi$ with $R_2 = 2$.

1.2 Linear Operators in Hilbert Spaces

This section is devoted to studying the different properties of linear operators between Hilbert spaces: their domains, ranges, norms, boundedness, closedness, and to examining special classes of operators: adjoint and self-adjoint operators, projections, isometric and unitary operators, functionals, time-evolution operators.

Great attention is paid to the notion of eigenvalues and eigenvectors (often said eigenfunctions, when the problem involves spaces of functions), due to its relevance in physical problems. Many exercises propose the different procedures needed for finding eigenvectors and the extremely various situations which can occur. According to the physicists use, the term "eigenvector" is used instead of the more correct "eigenspace", and "degeneracy" instead of "geometrical multiplicity" of the eigenvalue (i.e., the dimension of the eigenspace). The term "not degenerate" is also used instead of "degeneracy equal to 1". The notion of spectrum is only occasionally mentioned.

Another frequent question concerns the convergence of given sequences of operators in a Hilbert space H. Let us recall that the convergence of a sequence T_n to T as $n \to \infty$ is said to be
(i) "in norm" if $\|T_n - T\| \to 0$
(ii) "strong" if $\|(T_n - T)x\| \to 0,\ \forall x \in H$
(iii) "weak" if $\big(y, (T_n - T)x\big) \to 0,\ \forall x, y \in H$

Clearly, norm convergence implies strong and strong implies weak convergence, but the converse is not true. Many of the exercises provide several examples of this. Similar definitions hold for families of operators T_a depending on some continuous parameter a.

Questions as "Study the convergence" or "Find the limit" of the given sequence T_n (or family T_a) of operators are actually "cumulative" questions, which indeed include and summarize several aspects. A first aspect is to emphasize the fact that "convergence" (and the related notions of "approximation" and "neighborhood") is a "relative" notion, being strictly dependent on the definition of convergence which has been chosen. The next "operative" aspects are that, given the sequence of operators, one has to
(a) conjecture the possible limit T (this is usually rather easy)
(b) evaluate some norms of operators $\|T_n - T\|$ and/or of vectors $\|(T_n - T)x\|$ and so on, in order to decide what type of convergence is involved.

Frequent use will be done in this section, and also in Chap. 3, of the Lebesgue dominated convergence theorem *(briefly: Lebesgue theorem) concerning the convergence of integrals of sequences of functions. The statement of the theorem in a form convenient for our purposes is the following:*

Assume that a sequence of real functions $\{f_n(x)\} \in L^1(\mathbf{R})$ satisfies the following hypotheses
(i) $f_n(x)$ converges pointwise almost everywhere to a function $f(x)$,
(ii) there is a function $g(x) \in L^1(\mathbf{R})$ such that

$$|f_n(x)| \le g(x)$$

then:
(a) $f(x) \in L^1(\mathbf{R})$,
(b) $\displaystyle\lim_{n\to\infty} \int_{-\infty}^{+\infty} f_n(x)\,dx = \int_{-\infty}^{+\infty} f(x)\,dx$

The theorem is clearly also true if, instead of a sequence of functions depending on a integer index n, one deals with family of functions $\{f_a(x)\}$ depending on a continuous parameter a, and—typically—one considers the limit as $a \to 0$. See Problem 1.1 for simple examples of sequences of functions not satisfying the assumptions of this theorem.

Other examples of linear operators will be proposed in Sects. 3.1.2 and 3.2.2 in the context of Fourier transforms.

1.2.1 A Survey of General Properties of Operators

1.42

Assume for simplicity that the operators considered in this problem are bounded, although it could be easily seen that many of the properties listed below are shared also by unbounded operators, under the assumption that their domain is dense in the Hilbert space (this ensures the possibility of defining the adjoint operator T^+).

(1)*(a)* Let $V \subset H$ be an invariant subspace under an operator T in a Hilbert space H, i.e., $T : V \to V$. Show that $T^+ : V^\perp \to V^\perp$, where $V^\perp$ is the orthogonal complementary subspace to V.

(b) Show that $\mathtt{Ker}\,(T^+) = \big(\mathtt{Ran}\,(T)\big)^\perp$ and $\mathtt{Ker}\,(T) = \big(\mathtt{Ran}\,(\mathtt{T}^+)\big)^\perp$. What changes if T is unbounded?

(2)*(a)* Show that if T admits an eigenvector v, it is *not* true in general that v is also eigenvector of T^+ (a counterexample where T is a 2×2 matrix is enough).

(b) Assume that T is normal, i.e., $TT^+ = T^+T$, then show:

(i) If T admits an eigenvector v, then v is also eigenvector of T^+ (with eigenvalue . . .). *Hint*: start with $\|(T - \lambda I)v\| = 0$.

(ii) If T admits two eigenvectors with different eigenvalues, then these eigenvectors are orthogonal.

(3)*(a)* Show that the eigenvalues (if any) of a Hermitian operator are real

(b) Show that if an operator T admits a complete set of orthonormal eigenvectors with real eigenvalues λ_n, then it is Hermitian

(4) Assume that T admits a complete set of orthonormal eigenvectors with eigenvalues λ_n, then:
(a) Show that $\|T\| = \sup_n |\lambda_n|$. Show that this is no longer true if the eigenvectors are not a complete set or are not orthogonal (counterexamples where T are 2×2 matrices are enough, see also various problems in this book)
(b) Show that $TT^+ = T^+T$. What are the eigenvalues of TT^+?

(5) An operator A is called *positive* if $(x, Ax) \geq 0,\ \forall x \in H$; show that its eigenvalues λ (if any) satisfy $\lambda \geq 0$. Show that, given any T, then TT^+ and T^+T are positive, and that if for some $x \in H$ one has $T^+Tx = 0$, then also $Tx = T^+x = 0$.

1.43

(1) Show that an "isometric" operator T defined in a Hilbert space H, i.e., an operator preserving norms, is injective and admits only trivial kernel: $\mathtt{Ker}\,T = \{0\}$. What about its eigenvalues (if any)? Assume that T admits a complete set of orthonormal eigenvectors with eigenvalues λ_n: show that T is unitary if $|\lambda_n| = 1$.

(2) Let U be an unitary operator in H: show that a set $\{e_n\} \in H$ is orthonormal complete if and only if the same holds for $\{U\, e_n\}$. Conversely, show that if an operator $T : H \to H$ maps an orthonormal complete system $\{e_n\}$ into an orthonormal complete system $\{v_n\}$, then it is unitary (show first that the domain and the range of

T coincide with H, next that $(Tx, Ty) = (x, y)$, $\forall x, y \in H$, or—equivalently but more simply, as well-known—that $\|Tx\| = \|x\|$).

(3) Let $x \in H$: is the series, where $v_n = T\, e_n$ as in (2),

$$\sum_n (e_n, x)\, v_n$$

convergent in H? to what vector?

(4) Let U be the unitary operator which maps the "canonical" orthonormal complete system[2] $\{e_n,\ n = 1, 2, \ldots\}$ in ℓ^2 into the orthonormal complete system w_n defined in Problem 1.16. Find $U^{-1}e_n$.

(5) Let S be an invertible (in particular, unitary) operator, and let T be any operator admitting an eigenvector: $Tv = \lambda v$; show that $w = Sv$ is eigenvector of $T_S = STS^{-1}$.

1.44

(1) Let T and S two commuting operators: $[T, S] = TS - ST = 0$. Assume that T admits an eigenvalue λ with *finite* degeneracy n. What can be deduced about the existence of eigenvectors of S? And if the degeneracy of λ is infinite?

(2) *(a)* Assume that T and S satisfy ("ladder operators")

$$[T, S] = \sigma S\,, \qquad \sigma \in \mathbf{C}$$

and let $v \in H$ be an eigenvector of T with eigenvalue λ. Show that $Sv,\ S^2v, \ldots$ are eigenvectors of T with eigenvalue ...

(b) Deduce from *(a)* $[T^+, S^+] = \cdots$; extend then the result seen in *(a)* to the cases of T Hermitian and of T normal (see Problem 1.42 q. (2) *(b)* *(i)*)

1.45

Let $H_1,\ H_2$ be two Hilbert subspaces of a Hilbert space and let $P_1,\ P_2$ be the corresponding projections.

(1) Under what condition on $P_1,\ P_2$ is $T = P_1 + P_2$ a projection? on what subspace? Give an example.

(2) Under what condition on $P_1,\ P_2$ is $T = P_1 P_2$ a projection? Give an example. *Hint*: it is enough to impose self-adjointness ...

1.46

(1) What condition ensures that a projection is a compact operator?

(2) Is it true that an operator having finite-dimensional range is compact? is bounded?

(3) Show that any bounded operator B maps weakly convergent sequences of vectors into weakly convergent sequences. If B is bounded and C is compact, is it true that BC and CB are compact operators?

---

[2] I.e., $e_1 = (1, 0, 0, \ldots)$, $e_2 = (0, 1, 0, 0, \ldots)$, etc.

1.47

Let T be a bounded operator in a Hilbert space admitting a "discrete" spectral decomposition

$$T = \sum_{n=1}^{\infty} \lambda_n P_n$$

with standard notations. Show that the series $\sum_{n=1}^{\infty} \lambda_n P_n$ (more correctly: the partial sum $T_N = \sum_{n=1}^{N} \lambda_n P_n$) is norm-convergent if the eigenvalues satisfy $|\lambda_n| \to 0$, and strongly convergent if λ_n are bounded: $|\lambda_n| < M$.

1.48

(1) Let A be an operator in a Hilbert space H and consider the bilinear form defined by

$$< u, v >= (u, Av) , \qquad u, v \in H$$

Under what conditions on the operator A does this linear form define a scalar product in H?

(2) Assume that A, in addition to the conditions established in (1), is a bounded operator: show that if a sequence of vectors u_n is a Cauchy sequence with respect to the norm induced by the usual scalar product (,), then it is also a Cauchy sequence with respect to the norm induced by the scalar product defined by the bilinear form $< \, , \, >$ given in (1). Is the converse true? What changes if A is unbounded?

1.2.2 *Linear Operators Defined Giving $T\, e_n = v_n$, and Related Problems*

A common and convenient way to define a linear operator is that of assigning the results obtained when it is applied to an orthonormal complete system $\{e_n\}$ in a Hilbert space, i.e. of giving $v_n = T e_n$. Some significant cases are proposed in this subsection; other examples can also be found in the following subsections.

The first problem is to check if the domain of these linear operators can be extended to the whole Hilbert space in such a way to obtain a continuous operator (whenever possible; actually, the question "find $\|T\|$" always implies this preliminary step). Let us start with the simplest cases in the two following problems, where the $\{e_n\}$ are eigenvectors of T.

1.49

Let $\{e_n\}$ be an orthonormal complete system in a Hilbert space H, and let T be operators of the form

$$T\, e_n = c_n e_n\,, \qquad c_n \in \mathbf{C};\ \text{ no sum over } n$$

For each one of the cases listed below:

(a) Find the degeneracy of the eigenvalues, find $\|T\|$ (and specify if there is some $x_0 \in H$ such that $\|Tx_0\| = \|T\|\|x_0\|$), or show that T is unbounded

(b) Find domain and range of T (check in particular if they coincide with the whole space H or—at least—if they are dense in it)

(1) Let $n = 1, 2, \ldots$:

(i) $c_n = 1/n$; (ii) $c_n = n$; (iii); $c_n = \frac{n-i}{n+i}$

(2) Let $n \in \mathbf{Z}$:

(i) $c_n = \exp(in\pi/7)$; (ii) $c_n = \exp(in)$; (iii) $c_n = \frac{n^2}{2n+1}$; (iv) $c_n = \frac{n^2-1}{n^2+1}$

1.50

Let $\{e_n\,, n \in \mathbf{Z}\}$ and let T_N (where N is a fixed integer) be defined by

$$T_N\, e_n = e_n \quad \text{for} \quad |n| \le N \quad \text{and} \quad T_N\, e_n = 0 \quad \text{for} \quad |n| > N$$

(1) Show that T_N is a projection.[3] Is it compact?

(2) Study the convergence as $N \to \infty$ of the sequence of operators T_N to the operator $T_\infty =$ the identity I.

1.51

Let $\{e_n\,, n \in \mathbf{Z}\}$ and let S_N (where N is a fixed integer) be defined by

$$S_N\, e_n = e_{-n} \text{ for } 1 \le |n| \le N, \quad S_N\, e_0 = e_0 \text{ and } S_N\, e_n = 0 \text{ for } |n| > N$$

(1) Find the eigenvectors and eigenvalues (with their degeneracy) of S_N. Is it compact?

(2) Consider the operator S_∞ defined by $S_\infty\, e_n = e_{-n}$ for all nonzero $n \in \mathbf{Z}$ and with $S_\infty e_0 = e_0$. Study the convergence as $N \to \infty$ of the sequence of operators S_N to the operator S_∞.

(3) If $e_n = \exp(inx)/\sqrt{2\pi}$ in $H = L^2(-\pi, \pi)$ show that the operator S_∞ takes a very simple form!

[3] In this book, only *orthogonal* projections P will be considered, i.e. operators satisfying the properties $P^2 = P$ (idempotency) and $P^+ = P$ (Hermiticity).

1.52
In a Hilbert space H with orthonormal complete system $\{e_n, n = 1, 2, \ldots\}$ consider the set of vectors

$$v_1 = \frac{e_1 + e_2}{\sqrt{2}}, \ v_2 = \frac{e_3 + e_4}{\sqrt{2}}, \ldots, v_n = \frac{e_{2n-1} + e_{2n}}{\sqrt{2}}, \ldots, \qquad n = 1, 2, \ldots$$

(1) Is $\{v_n\}$ a orthonormal set? a complete set?

(2) Let T be the linear operator defined by

$$T\, e_n = v_n$$

(a) Show that $\mathtt{Ran}\, T$ is a Hilbert subspace of H; what is its orthogonal complementary subspace and the dimension of this subspace?

(b) Does T preserve scalar products? is it unitary?

(c) Does T admit eigenvectors? What is its kernel? (recall the properties of isometric operators, see Problem 1.43, q. (1)).

1.53
In a Hilbert space H with orthonormal complete system $\{e_n,\ n = 1, 2, \ldots\}$ consider the linear operators ("shift operators") defined by

$$T\, e_n = e_{n+1} \quad \text{and} \quad \begin{cases} S\, e_1 = 0 \\ S\, e_n = e_{n-1} \text{ for } n > 1 \end{cases}$$

(1) Writing a generic vector $x \in H$ in the form $x = \sum_n a_n\, e_n \equiv (a_1, a_2, \ldots)$ (equivalently: choose $H = \ell^2$), obtain Tx and Sx. Show that the domain of these operators is the whole Hilbert space.

(2) Is T injective? surjective? the same questions for S.

(3) Calculate $\|T\|$ and $\|S\|$.

(4) Show that $S = T^+$.

(5) Show that T is isometric, i.e., preserves scalar products: $(x, y) = (Tx, Ty)$, $\forall x, y \in H$ but is not unitary. Study the operators TT^+ and T^+T. Show that TT^+ is a projection: on what subspace?

(6) Show that T has no eigenvectors (as in previous problem, recall the properties of isometric operators), but $S = T^+$ has many (!) eigenvectors.

(7) Are T and S compact operators?

1.54

(1) What changes in the above problem if $T\, e_n = e_{n+1}$ is defined on an orthonormal complete system where now $n \in \mathbf{Z}$?

(2) A "concrete" version of this operator is the following: let $H = L^2(0, 2\pi)$ with $e_n = \exp(inx)/\sqrt{2\pi}$. Then T becomes simply $T\, f(x) = \exp(ix) f(x)$ for any $f \in L^2(0, 2\pi)$. Find again (and confirm) the results obtained before.

1.55
Although the operators considered in this problem are not of the form $Te_n = v_n$ which is that considered in this subsection, they share similar properties with those of Problems 1.53 and 1.54; in particular they are resp. an isometric and an unitary operator.

(1) In $H = L^2(0, +\infty)$ consider $T\, f(x) = f(x-1)$; to avoid difficulties with the restriction $x > 0$, it is convenient to write for clarity

$$T\, f(x) = f(x-1)\theta(x-1) \quad \text{and} \quad S\, f(x) = f(x+1)\theta(x)$$

where

$$\theta(x) = \begin{cases} 0 \text{ for } x < 0 \\ 1 \text{ for } x > 0 \end{cases}$$

The same questions (2–7) as in Problem 1.53. For what concerns the eigenvectors of S, consider only the functions $g(x)$ which assume a constant value c_n on each interval $n < x < n+1$, and the functions $h(x) = \exp(-\alpha x)$, $\alpha > 0$. Find also $\mathtt{Ker}\, S$.
(2) What changes if $T\, f(x) = f(x-1)$ is defined in $L^2(\mathbf{R})$? (compare with Problem 1.54, q.(1))

1.56
Another operator with similar properties as T in Problem 1.53:

In a Hilbert space H with orthonormal complete system $\{e_n,\ n = 1, 2, \ldots\}$ consider the operators

$$T\, e_n = e_{2n} \quad \text{and} \quad \begin{cases} S\, e_n = e_{n/2} \text{ for } n \text{ even} \\ S\, e_n = 0 \quad \text{for } n \text{ odd} \end{cases}$$

Exactly all the same questions (1–7) as in Problem 1.53. Show that S has many and many eigenvectors.

1.57
The same remark as for Problem 1.55. Another case:

Consider the operators T and S of $L^2(0, 1)$ in itself defined resp. by

$$T\, f(x) = g(x) \ \text{ where } g(x) = f(2x)$$

(warning: $f(x)$ is given only in $0 < x < 1$, so it defines $g(x)$ only if $0 < x < 1/2$; it is understood that $g(x)$ is put equal to zero if $1/2 < x < 1$); and let S be defined by

$$S\,f(x) = h(x) \text{ where } h(x) = (1/2)f(x/2)$$

The same questions (2–7) as in Problem 1.53. For what concerns the eigenvectors of S, consider only the functions $f(x) = x^\alpha$. Verify that the corresponding eigenvalues λ_α satisfy the condition $\|S\| \geq \sup_\alpha |\lambda_\alpha|$.

1.58

Study the convergence as $N \to \infty$ of the sequences of operators T^N, S^N where T and S are some of the isometric and unitary operators considered in the previous Problems 1.53–1.55. Precisely:

(1) if T and S are given in Problem 1.55, q.(1), i.e. $T^N f(x) = f(x-N)\theta(x-N)$ in $L^2(0, +\infty)$, etc. for S^N.

(2) if T and S are given in Problem 1.55, q.(2), i.e. the same as in (1) but in $L^2(\mathbf{R})$ (use Fourier transform; see also Problem 3.20).

(3) if T and S are given in Problem 1.53, i.e. $T^N e_n = e_{n+N}$ etc. with $n = 1, 2, \ldots$.

(4) if T and S are given in Problem 1.54, q.(2), i.e. $T^N f(x) = \exp(iNx)f(x)$ in $L^2(0, 2\pi)$.

(5) if T and S are given in Problem 1.54, q.(1), i.e. the same as in (3): $T^N e_n = e_{n+N}$ but with $n \in \mathbf{Z}$.

1.59

(1) Let $v_n = e_n - e_{n-1}$, where $\{e_n,\ n \in \mathbf{Z}\}$ is an orthonormal complete system in a Hilbert space H and let T be the operator

$$T\,e_n = v_n$$

(a) find $\mathtt{Ker}\,T$
(b) show that $\|T\| \leq 2$
(c) find $T(e_0 + e_1 + \cdots + e_k)$: what information can be deduced about the boundedness of T^{-1}?

(2) More in general, let $w_n = \alpha e_n - \beta e_{n-1}$ with nonzero $\alpha, \beta \in \mathbf{C}$, and let

$$T\,e_n = w_n$$

(a) is the set w_n a complete set in H? deduce: is $\mathtt{Ran}\,T$ a dense subspace in H?
(b) look for eigenvectors of T.

(3) If now $H = L^2(0, 2\pi)$ and $e_n = \exp(inx)/\sqrt{2\pi}$, the above operator acquires a concrete (possibly simpler) form: $Tf(x) = \varphi(x)f(x)$ where $f(x) \in L^2(0, 2\pi)$ and $\varphi(x) = \cdots$.
(a) find $\|T\|$
(b) confirm the results seen in (2)
(c) under what conditions about α, β does the operator T given in (2) admit bounded inverse?

(d) in the case $\alpha = \beta = 1$, does the function $g_1(x) = 1$ belong to $\mathtt{Ran}\, T$? and the function $g_2(x) = \sin x$?

1.60
Consider the operator defined in a Hilbert space H with orthonormal complete system $\{e_n,\ n = 1, 2, \dots\}$

$$T\, e_n \;=\; c_n x_0 + e_n\,, \qquad\qquad n = 1, 2, \dots$$

where $\{c_n\} \in \ell^2,\ c_n \in \mathbf{R}$ and $x_0 \in H$ are given.

(1) Show that T is bounded.

(2) Find eigenvalues and eigenvectors of T. What condition on c_n and x_0 ensures that the eigenvectors provide a complete set for H?

(3) In the case $x_0 = \sum_m c_m e_m$, find $\|T\|$ and check if the eigenvectors provide a complete set for H.

1.61
Let $\{e_n,\ n \in \mathbf{Z}\}$ be an orthonormal complete system in a Hilbert space H and let T be the operator defined by

$$T\, e_n \;=\; \alpha_n e_{n+1}\,, \qquad\qquad \alpha_n \in \mathbf{C}$$

(1) For what choice of α_n:
(a) is T unitary? *(b)* is T bounded? *(c)* is T a projection?

(2) Find T^+.

(3) Let $\alpha_n = \exp(in\pi/2) + i$:
(a) find $\|T\|$
(b) find $\mathtt{Ker}\, T$; are there vectors $\in \mathtt{Ker}\, T$ which also belong to $\mathtt{Ran}\, T$?

1.62
Let $\{e_n,\ n = 1, 2, \dots\}$ be an orthonormal complete system in a Hilbert space H and let T the operator defined by

$$T\, e_n \;=\; x_0$$

where x_0 is a *fixed* nonzero vector.

(1) *(a)* is the domain of T the whole Hilbert space?

(b) is T a bounded operator?

(2) What is the kernel of T? (see Problem 1.16)

(3) Construct two sequences of vectors z_n and w_n both tending as $n \to \infty$ to e_1 (e.g.) but such that $T(z_n) = 0$ and $T(w_n) \to e_1$.

(4) Conclude: is T a closed operator? Show that the kernel of any closed operator is a closed subspace.

1.63

Consider the operator defined in a Hilbert space H with orthonormal complete system $\{e_n,\ n = 1, 2, \ldots\}$

$$T\,e_n = \alpha e_n + \beta e_1\,, \qquad \alpha, \beta \in \mathbf{C};\ \alpha, \beta \neq 0$$

(1) The same questions as in (1) of the previous problem.

(2) For what values of α, β does T admit a nontrivial kernel? of what dimension?

(3) Show that T coincides with a multiple of the identity operator in a dense subspace of H.

(4) Is T a closed operator?

1.64

Let $\{e_n,\ n \in \mathbf{Z}\}$ be an orthonormal complete system in a Hilbert space H and let T be the operator defined by

$$T\,e_0 = c_0 e_0 \quad , \quad T\,e_n = c_n e_{-n}\,, \qquad n \neq 0,\ c_n \in \mathbf{C}$$

(1) Give the conditions on the coefficients c_n in order to have
(*a*) T bounded; (*b*) T normal, i.e., $TT^+ = T^+T$; (*c*) T Hermitian; (*d*) $T^2 = I$

(2) Show that the problem of looking for the eigenvectors of T reduces to simple problems in 2–dimensional subspaces (or to the trivial one-dimensional case $T\,e_0 = c_0 e_0$). Is the set of the eigenvectors a complete set in H?

(3) Consider the particular cases:
(*a*) $c_n = -c_{-n} \neq 0$: find the eigenvectors and eigenvalues (with their degeneracy): can one expect that the eigenvectors are orthogonal? and the eigenvalues real?
(*b*) $c_n = \begin{cases} n & \text{for} \quad n > 0 \\ 1/n^2 & \text{for} \quad n < 0 \end{cases}$: find the eigenvectors and eigenvalues (with their degeneracy); the sequence of the eigenvalues is bounded and converges to 0 as $n \to \infty$, however T is not compact (actually, it is unbounded), is this surprising?

1.65

Let $\{e_n = \exp(inx)(2\pi)^{-1/2},\ n \in \mathbf{Z}\}$ in the space $L^2(-\pi, \pi)$, and let T be defined by

$$T\,e_0 = 0 \quad , \quad T\,e_n = n^2 e_{-n}$$

(1) Find eigenvectors and eigenvalues of T with their degeneracy. Do the eigenvectors provide a complete set for the space?

(2) For what values of $c \in \mathbf{C}$ is the operator $T + cI$ invertible?

(3) Find $\|(T + 20\,I)^{-1}\|$, $\|(T + i\,I)^{-1}\|$, $\|\big(T + (2+i)\big)^{-1}\|$

1.66

Let $e_n = \exp(inx)(2\pi)^{-1/2}$, $n \in \mathbf{Z}$ in the space $L^2(-\pi, \pi)$, and let T be defined by

$$T\,e_0 = 0 \;,\quad T\,e_n = \frac{1}{n^2}e_{-n} \quad \text{for} \quad n \neq 0$$

(1) Find eigenvectors and eigenvalues of T with their degeneracy.
(2) What is $\mathtt{Ran}\,T$? Specify if it is a Hilbert subspace of $L^2(-\pi, \pi)$, or—alternatively—what is its closure.
Consider now the equation

$$T\,f = g$$

where $g = g(x) \in L^2(-\pi, \pi)$ is given and $f = f(x) \in L^2(-\pi, \pi)$ unknown.
(3) *(a)* Let $g(x) = \cos^2(x) - (1/2)$: does this equation admit solution? is the solution unique?
(b) Same questions if $g(x) = \cos^4(x)$ (no calculation needed!).
(c) Same questions if $g(x) = |x| - \pi/2$ (no calculation needed!).
(4) Show that the sequence of operators T^N is norm-convergent as $N \to \infty$: to what operator?

1.67

Let $\{e_n,\ n = 1, 2, \ldots\}$ be an orthonormal complete system in a Hilbert space H and let T be the operator defined by

$$T\,e_n = \alpha_n e_1 + \beta_n e_2$$

where $n = 1, 2, \ldots$ and α_n and β_n are given sequences of complex numbers.

(1) Let $\alpha_n = \beta_n = 1/2^n$. *Hint*: put $z = \sum_n e_n/2^n$, then:
(a) find $\|T\|$, *(b)* find eigenvectors and eigenvalues of T, *(c)* find T^+
(2) For what α_n and β_n is T bounded?
(3) *(a)* For what α_n and β_n is the range of T 1-dimensional?
(b) For what α_n and β_n are the range and the kernel of T orthogonal?
(4) Fixed an integer $N > 1$, let $\alpha_n = \beta_n = \begin{cases} 0 & \text{for } n \le N \\ 1/2^{(n-N)} & \text{for } n > N \end{cases}$ and let T_N be the corresponding operator. Study the convergence as $N \to \infty$ of the sequences of operators T_N and T_N^+. *Hint*: put $z_N = \sum_{n>N}^{\infty} e_n/2^{(n-N)}$, then $T_N x = \cdots$

1.68

Let $\{e_n\ , n \in \mathbf{Z}\}$ be an orthonormal complete system in a Hilbert space H and, fixed any integer N, consider the operators defined by

$$S\,e_n = \frac{n^2}{1+n^4}e_n \;,\quad T_N\,e_n = \begin{cases} e_n & \text{for} \quad |n| \le N \\ e_{-n} & \text{for} \quad |n| > N \end{cases}$$

(1) Find eigenvectors and eigenvalues (with their degeneracy) of S and of T_N. Is there an orthogonal complete set for H of simultaneous eigenvectors of S and T_N?

(2) Is the operator S compact? and $T_N\, S$?

(3) Study the convergence of the sequence of operators T_N and $T_N S$ as $N \to +\infty$.

1.2.3 Operators of the Form $T\,x = v(w, x)$ and $T\,x = \sum_n v_n(w_n, x)$

It can be noted that the operators of the previous subsection can be viewed as a special case of the operators considered here. Indeed, choosing $w_n = e_n$, where $\{e_n\}$ is a orthonormal complete system in the Hilbert space, one obtains just $T\,e_n = v_n$.

1.69

Consider in $L^2(-a, a)$ $(a > 0, \neq \infty)$ the operator

$$T\,f(x) = h(x) \int_{-a}^{a} f(x)\,dx$$

where $h(x) \in L^2(-a, a)$ is a given function.

(1) Find the domain, the kernel and the range of T, with their dimensions.

(2) Find the eigenvectors and eigenvalues of T with their degeneracy.

(3) Find $\|T\|$.

(4) Find the adjoint operator T^+, its eigenvectors and eigenvalues with their degeneracy.

(5) Study the operator T^2.

1.70

The same as the above problem but with $a = \infty$, i.e.

$$T\,f(x) = h(x) \int_{-\infty}^{+\infty} f(x)\,dx$$

(1) Find eigenvectors and eigenvalues of T with their degeneracy (distinguish the cases $h(x) \in L^1(\mathbf{R}) \cap L^2(\mathbf{R})$ and $h(x) \notin L^2(\mathbf{R})$ or $h \notin L^1(\mathbf{R})$)

(2) Specify the domain and the kernel of T (see also Problem 1.4).

(3) Show that T is not closed (see also Problem 1.4).

(4) Show that if the kernel of an operator is *dense* in the Hilbert space, then the operator is not closed (cf., for a different example, Problem 1.62).

1.71

(1) Let I be an interval $I \subseteq \mathbf{R}$ and consider the operator

$$T f(x) = h(x) \int_I g(x) f(x) \, dx = h(g^*, f)$$

where $h(x)$ and $g(x)$ are given functions in $L^2(I)$. The same questions (1–5) as in Problem 1.69.

(2) The operator in (1) can be generalized in abstract setting in the form

$$T x = v(w, x)$$

where v, w are given vectors in a Hilbert space.

(a) The same questions (1–5) as in Problem 1.69.

(b) Is it possible to choose v, w in such a way that T is a projection?

(c) Show that it is possible to choose v, w in such a way that $T^2 = 0$ (but $T \neq 0$).

1.72

(1) For each fixed integer N consider the operators in $L^2(0, 2\pi)$

$$T_N f(x) = \frac{1}{2\pi} \sum_{n=-N}^{N} \exp(inx) \int_0^{2\pi} \exp(-iny) f(y) \, dy$$

Recognize that these operators admit elementary properties ...; study the convergence as $N \to \infty$ of the sequence of the operators T_N; see the following (2)(a).

(2) Consider now the operator

$$C f(x) = \frac{1}{2\pi} \sum_{n \in \mathbf{Z}} c_n \int_0^{2\pi} \exp\left(in(x - y)\right) f(y) \, dy$$

(a) let $c_n = 1$: then C becomes a trivial operator ...
(b) and if $c_n = in$?
(c) let instead $c_n = 1/2^{|n|}$: show that the functions $g(x) = Cf(x)$ in $\mathtt{Ran}\, C$ have special continuity properties.

1.73

Consider the following (apparently similar) operators defined in $L^2(0, 2\pi)$:

$$A_n^{(\pm)} f(x) = \frac{1}{2\pi}\int_0^{2\pi} \exp\big(in(x \pm y)\big) f(y)\, dy\,, \qquad n \in \mathbf{Z}$$

$$B_n^{(\pm)} f(x) = \frac{1}{\pi}\int_0^{2\pi} \sin\big(n(x \pm y)\big) f(y)\, dy\,, \qquad n = 1, 2, \ldots$$

$$C_n^{(\pm)} f(x) = \frac{1}{\pi}\int_0^{2\pi} \cos\big(n(x \pm y)\big) f(y)\, dy\,, \qquad n = 1, 2, \ldots$$

(1) What of these operators are projections? on what subspace?

(2) Find eigenvectors and eigenvalues with their degeneracy of the above operators.

(3) Study the convergence of the sequences of these operators as $n \to \infty$.

1.74

In the space $L^2(-\pi, \pi)$ consider the operators

$$T_N f(x) = \frac{1}{\pi}\sin x \int_{-\pi}^{\pi} \sin y\, f(y)\, dy + \frac{1}{\pi}\sin(2x)\int_{-\pi}^{\pi} \sin(2y)\, f(y)\, dy + \cdots$$
$$+ \frac{1}{\pi}\sin(Nx)\int_{-\pi}^{\pi} \sin(Ny)\, f(y)\, dy$$

where $N \geq 1$ is an integer.

(1) Show that T_N is a projection, find its eigenvectors and eigenvalues with their degeneracy.

(2) Study the convergence as $N \to +\infty$ of the sequence of the operators T_N and show that also the limit operator T_∞ is a projection.

(3) Are the operators T_N compact operators? and the operator T_∞?

(4) A "variation" (and an abstract version) of this problem: in a Hilbert space H with orthonormal complete system $\{e_n,\ n \in \mathbf{Z}\}$, consider the operators

$$T_N x = \sum_{n=-N}^{N} e_n (e_n, x)\,, \qquad x \in H$$

The same questions (1)–(3). What is in this case the operator T_∞? Compare with the Problem 1.50.

1.75

This problem looks at first sight quite similar to the previous one. This is not the case (what is the main difference?): let $H = L^2(-1, 1)$ and let T_N be defined by (with $N \geq 1$)

$$T_N f(x) = \int_{-1}^{1} f(y)\,dy + x\int_{-1}^{1} y\, f(y)\,dy + \cdots + x^N \int_{-1}^{1} y^N f(y)\,dy$$

(1) Find $\mathtt{Ran}\, T_N$ and $\mathtt{Ker}\, T_N$ with their dimensions.

(2) Fix $N = 2$: are the functions $f_0(x) = 1,\ f_1(x) = x,\ f_2(x) = x^2,\ f_3(x) = x^3$ eigenfunctions of T_2?

(3) Is T_2 (and in general T_N) a projection?

(4) Find the eigenvectors and eigenvalues of T_2.

1.76

For each integer N let $\chi_N(x)$ be the characteristic function of the interval $(-N, N)$ and let T_N be the operators in $L^2(\mathbf{R})$

$$T_N\, f(x) = \chi_N(x) \int_{-N}^{N} f(x)\,dx\,, \qquad N = 1, 2, \ldots$$

(1) Find T_N^2. Show that T_N is a projection apart from a factor c_N.

(2) Show that $T_M T_N \neq T_N T_M$ (if $N \neq M$, of course).

(3) Find eigenvectors and eigenvalues (with their degeneracy) of $T_1 T_N$ and of $T_N T_1$.

(4) Are there *even* functions $f(x) \in L^2(\mathbf{R})$ such that $T_N(f) = 0$ for all N?

1.77

In the Hilbert space $L^2(0, +\infty)$ let $\chi_n(x)$ be the characteristic function of the interval $(n-1, n)$, $n = 1, 2, \ldots$

(1) Show that, for any $f \in L^2(0, +\infty)$, the sequence $c_n = (\chi_n, f) \in \ell^2$.

(2) Consider the operators, for each integer N,

$$T_N\, f(x) = \sum_{n=1}^{N} \chi_n(x) \int_0^{+\infty} \chi_n(y) f(y)\,dy = \sum_{n=1}^{N} \chi_n(x) \int_n^{n+1} f(y)\,dy$$

Fixed $N \geq 1$, find eigenvectors and eigenvalues of T_N, with their degeneracy.

(3) Consider the operator

$$T_\infty\, f(x) = \sum_{n=1}^{\infty} \chi_n(x) \int_0^{+\infty} \chi_n(y) f(y)\,dy = \sum_{n=1}^{\infty} \chi_n(x) \int_n^{n+1} f(y)\,dy$$

(a) is T_∞ defined in the whole space?
(b) find its norm.
(c) is it compact?

(4) Study the convergence as $N \to \infty$ of the sequence of operators T_N to T_∞.

(5) Consider now

$$S_N f = \sum_{n=1}^{N} \frac{1}{n} \chi_n(x) \int_0^{+\infty} \chi_n(y) f(y)\, dy$$

(a) study the convergence as $N \to \infty$ of the sequence of operators S_N to S_∞.
(b) is S_∞ compact?

1.78

Let T be the operator in $L^2(-\pi, \pi)$

$$T f(x) = \sum_{n=1}^{\infty} \frac{a_n}{n} \sin nx \quad \text{where} \quad a_n = \frac{1}{\pi} \int_{-\pi}^{\pi} f(x) \sin nx\, dx$$

(1) Find $T(f)$ if $f(x) = \exp(2ix)$

(2) Find eigenvectors and eigenvalues of T with their degeneracy. Is T compact?

(3) Is it possible to write T in the form $T = \sum_n \lambda_n P_n$, i.e., as a "discrete" spectral decomposition?

(4) Study $\mathtt{Ran}\ T$; clearly, it is contained in the Hilbert subspace of *odd* functions: does it coincide with this subspace or—at least—is dense in it? Is it true that if $g(x) \in \mathtt{Ran}\ T$ then $g(x)$ is a continuous function? Is the converse true?

1.79

In the space $L^2(-\pi, \pi)$ consider the operator

$$T f(x) = \sum_{n=1}^{\infty} \frac{1}{n\pi} \cos nx (\sin nx, f)$$

(1) Find $\|T\|$ and specify if there is some f_0 such that $\|T f_0\| = \|T\| \|f_0\|$.

(2) Find eigenvalues and eigenvectors with their degeneracy.

(3) For each $f(x) \in L^2(-\pi, \pi)$, is the function $g(x) = T f(x)$ a continuous function? Is its derivative a continuous function? a function in L^2? Is the operator $\frac{d}{dx} T$ a bounded operator?

(4) Does the equation (the unknown is $f(x)$)

$$T f(x) = \alpha + \beta x^3 + |x|$$

admit solution (it is not requested to obtain $f(x)$) for some values of the constants α, β?

1.80
In the space $L^2(0, \pi)$ consider, for each fixed $n = 1, 2, \ldots$, the operator

$$T_n f(x) = \cos nx \int_0^{\pi} \left(\cos ny - \frac{1}{\sqrt{2}} \right) f(y)\, dy$$

(1) Find $\|T_n\|$. *Hint*: introducing the orthonormal complete system $e_0 = 1/\sqrt{\pi}$, $e_n = \sqrt{2/\pi} \cos nx$, write the operator in a more convenient form ...

(2) Find $\mathrm{Ker}\ T_n$ and $\mathrm{Ran}\ T_n$: are they orthogonal?

(3) Find eigenvectors and eigenvalues of T_n. Is it true in this example that $\|T_n\| = \sup |\text{ eigenvalues } |$?

(4) Are there functions $f(x) \in L^2(0, \pi)$ such that $T_n(f) = 0$ for all n?

(5) Study the convergence as $n \to \infty$ of the sequence of operators T_n.

1.2.4 *Operators of the Form* $T\, f(x) = \varphi(x) f(x)$

1.81
(1) In the space $H = L^2(0, a)$ $(a \neq \infty)$ consider the operator

$$T\, f(x) = x\, f(x)$$

(a) Find $\|T\|$. Show that there is no function $f_0(x) \in H$ such that $\|T\, f_0\| = \|T\|\|f_0\|$, but construct a family of functions $f_\varepsilon(x) \in H$ such that $\sup_{\varepsilon \to 0} \|T\, f_\varepsilon\|/\|f_\varepsilon\| = \|T\|$

(2) Does T admit eigenvectors? Is its kernel trivial?

(3) Find the spectrum of T. (Recall: the *spectrum* of an operator T is the set of the numbers $\sigma \in \mathbf{C}$ such that $T - \sigma I$ does *not* admit bounded inverse, or—more explicitly—such that $T - \sigma I$ is either not invertible or admits *unbounded* inverse.)

1.82
Consider the same operator as before in the space $H = L^2(\mathbf{R})$:

(1) Is its domain the whole space H, or—at least—dense in it?

(2) Show that this operator is unbounded: construct a sequence of functions $f_n(x) \in H$ such that $\sup_n \|T\, f_n\|/\|f_n\| = +\infty$

(3) Is its range the whole space H, or—at least—dense in it?

(4) The same questions for the operators $T\, f(x) = x^a f(x)$ with $a > 0$.

1.83
Consider the operator in $L^2(\mathbf{R})$ with $\varphi = 1/x$ (which is the inverse of the operator of the previous problem), i.e.,

$$T\, f(x) = \frac{1}{x} f(x)\,, \qquad x \in \mathbf{R}$$

(1) Show that this operator is unbounded: construct a sequence of functions $f_n(x) \in H$ such that $\sup_n \|T\, f_n\|/\|f_n\| = +\infty$

(2) Study the domain and the range of this operator (cf. the previous problem)

1.84
In the space $H = L^2(\mathbf{R})$ consider the operator in the general form

$$T\, f(x) = \varphi(x)\, f(x)$$

where $\varphi(x)$ is a given (real or complex) function. Under what conditions on $\varphi(x)$:

(a) is T bounded? Then find its norm. And is there some function $f_0(x)$ such that $\|T\, f_0\| = \|T\|\, \|f_0\|$?
(b) is T Hermitian? is T normal? is T unitary? *(c)* is T a projection?
(d) is the domain of T dense in H (but $\neq H$)?
(e) does T admit bounded inverse? Then find $\|T^{-1}\|$
(f) is $\mathtt{Ran}\, T$ dense in H (but $\neq H$)?
(g) are there eigenfunctions of T? *(h)* is T compact?

1.85
Find kernel, domain and range, specifying if domain and range coincide with $H = L^2(\mathbf{R})$, or at least are dense in it, of the operators $Tf = \varphi f$ in each one of the following cases:

$$(a)\ \varphi = \frac{1}{1+x^2}\,, \quad (b)\ \varphi = \frac{x+i}{x-i}\,, \quad (c)\ \varphi = x + |x|\,,$$

$$(d)\ \varphi = \sin x\,, \quad (e)\ \varphi = \exp(-x^2)\,, \quad (f)\ \varphi = \exp(-1/x^2)$$

1.86
Consider the operator in $H = L^2(\mathbf{R})$ with $\varphi = x^2/(1+x^2)$, i.e.

$$T\, f(x) = \frac{x^2}{1+x^2} f(x)$$

(1) Find $\|T\|$. Is there some function $f_0(x) \in H$ such that $\|T\, f_0\| = \|T\|\,\|f_0\|$?

(2) Does T admit eigenvectors? Is it compact?

(3) Is $\mathtt{Ran}\, T$ coinciding with H or at least dense in it? And its domain?

(4) Study the convergence as $N \to +\infty$ of the sequence of operators T^N, i.e. of the operators $T^N f = \varphi^N f$.

1.87

The same questions as before for the operator $T f = \varphi f$ in $L^2(\mathbf{R})$ with

$$\varphi(x) = \begin{cases} 0 & \text{for } x < 0 \\ 1 & \text{for } 0 < x < 1 \\ 1/x & \text{for } x > 1 \end{cases}$$

In particular, is $\mathtt{Ran}\, T$ a Hilbert subspace of $L^2(\mathbf{R})$? is $\mathtt{Ran}\, T$ orthogonal to $\mathtt{Ker}\, T$? is it correct to say that $L^2(\mathbf{R}) = \mathtt{Ker}\, T \oplus \mathtt{Ran}\, T$?

1.88

(1) Consider in the space $H = L^2(0, 2\pi)$ the operator with $\varphi = \exp(ix)$, i.e.

$$T f(x) = \exp(ix) f(x)$$

For what values of $\rho \in \mathbf{C}$ does the operator $T - \rho I$ admit bounded inverse?

(2) *(a)* Find the norm of $(T + 2i\, I)^{-1}$

(b) Study the convergence of the sequence of operators T^N as $N \to +\infty$.

(3) The same questions as in (2) if $H = L^2(0, \pi)$.

(4) The same questions as in (2) if $H = L^2(\mathbf{R})$.

1.89

In the space $L^2(0, 2\pi)$ consider the operator

$$T f(x) = \big(1 - \alpha \exp(ix)\big) f(x), \qquad \alpha \in \mathbf{C}$$

(1) Find the adjoint T^+ and specify if T is normal.

(2) Find $\|T\|$. For what values of $\alpha \in \mathbf{C}$ does T admit bounded inverse?

(3) For what values of $\alpha \in \mathbf{C}$ is the series of functions

$$\sum_{n=0}^{\infty} \alpha^n \exp(inx)$$

convergent in $L^2(0, 2\pi)$? What is its limit? Is the convergence uniform?

(4) Consider now the operator $S = I - T$, i.e., the operator $S f(x) = \alpha \exp(ix) f(x)$: show that for $|\alpha| < 1$ the sequence A_N defined by

$$A_N = T \sum_{n=0}^{N} S^n = (I - S) \sum_{n=0}^{N} S^n$$

is norm-convergent to the identity, and then conclude

$$\sum_{n=0}^{\infty} S^n = (I - S)^{-1}$$

1.2.5 *Problems Involving Differential Operators*

1.90

(1) Consider in $L^2(0, 1)$ the operator $T = i\,d/dx$ defined in the domain D_0 of the functions $f(x)$ with vanishing boundary conditions $f(0) = f(1) = 0$ (the functions in the domain are clearly assumed to be differentiable in L^2). Find the adjoint of T and show that its domain D' is *larger* than D_0.

(2) Consider now the operator $T = i\,d/dx$ in the domain D_α of the functions with the boundary conditions $f(1) = \alpha f(0)$, with $\alpha \in \mathbf{C}$, $\alpha \neq 0$. For what values of α is T Hermitian?

(3) Find eigenvectors and eigenvalues of T (in D_α, for generic α). Are the eigenvectors orthogonal? Are the eigenvectors a complete set in $L^2(0, 1)$?

(4) Check the correctness of the answer given to (2) by comparison with the properties obtained in (3) (the eigenvectors are orthogonal and the eigenvalues real if ...)

(5) Are the all domains $D_0,\ D',\ D_\alpha$ dense in $L^2(0, 1)$?

For a discussion about existence and uniqueness of the solutions of the Problems 1.91–1.97, compare with Problems 1.32, 1.33 and in particular Problem *1.34. Clearly, the choice of x or t as independent variable is not relevant in the present context, and to avoid too frequent changes in the notations, we continue to use here x as independent variable.*

1.91

Consider in $H = L^2(-\pi, \pi)$ the operator, with $\alpha \in \mathbf{C}$, $\alpha \neq 0$,

$$T = \frac{d}{dx} + \alpha I \quad \text{with periodic boundary conditions : } f(-\pi) = f(\pi)$$

(1) Find eigenvalues and eigenvectors of T.

(2) Consider the equation $Tf = g$, i.e.

$$f' + \alpha f = g$$

where $g(x) \in H$ is given and $f(x) \in H$ the unknown. Expand $f(x)$ and $g(x)$ in Fourier series with respect to the eigenvectors found in (1). Under what condition on the constant α does this equation admit solution? In this case, is the solution unique?

(3) Let $\alpha = 2i$: under what condition on the the Fourier coefficients g_n of $g(x)$ does the equation admit solution? In this case, show that the solution is not unique and write as a Fourier series the most general solution.

(4) Under the condition found in (2), show that T is invertible and find $\|T^{-1}\|$ in each one of the cases

$$\alpha = 1, \quad \alpha = 2 + i, \quad \alpha = 1 + 2i, \quad \alpha = 3i/2$$

1.92
Consider in the space $H = L^2(0, \pi)$ the operator

$$T = \frac{d^2}{dx^2} \quad \text{with vanishing boundary conditions} : f(0) = f(\pi) = 0$$

(1) Show that this operator is Hermitian, i.e. $(g, Tf) = (Tg, f)$ in the dense domain of doubly differentiable functions $f(x),\ g(x) \in L^2$, satisfying these conditions.

(2) Solve the equation (where $g(x) \in H$ is given and $f(x) \in H$ the unknown)

$$\frac{d^2}{dx^2} f = g$$

expanding both $g(x)$ and $f(x)$ in Fourier series with respect to the complete set $\{\sin nx,\ n = 1, 2, \ldots\}$ of the eigenfunctions of T, and verify that the equation admits a unique solution.

(3) Compare with the next problem.

1.93
Consider in the space $H = L^2(0, \pi)$ the operator

$$T = \frac{d^2}{dx^2} \quad \text{with periodic boundary conditions, i.e., } f(0) = f(\pi),\ f'(0) = f'(\pi)$$

(1) The same question (1) as in the above problem, but now with periodic boundary conditions.

(2) Solve the equation ($g(x) \in H$ is given, $f(x) \in H$ the unknown)

$$\frac{d^2}{dx^2} f = g$$

now expanding both $g(x)$ and $f(x)$ in Fourier series with respect to the complete set $\{e_n = \exp(2inx),\ n \in \mathbf{Z}\}$ of the eigenfunctions of T (what is, in this case, the degeneracy of the eigenvalues?). What about the existence and uniqueness of the solution? What condition must be imposed to the function $g(x)$ in order that some solution exists?

1.94

In the space $H = L^2(0, \pi)$ consider the operator $T = -d^2/dx^2 + I$ and the equation $Tf = g$, i.e. ($g(x) \in H$ is given and $f(x) \in H$ the unknown)

$$-f'' + f = g$$

with vanishing boundary conditions $f(0) = f(\pi) = 0$. Expand $g(x)$ and $f(x)$ in Fourier series with respect to the orthogonal complete set $\{\sin nx,\ n = 1, 2, \ldots\}$.

(1) Show that the equation admits a unique solution for any $g(x) \in H$.

(2) Find $\|S\|$, where S is the inverse of the above operator T:

$$S g = f$$

(3) Let $g(x) \in H$ and let g_N be a sequence of functions "approximating" $g(x)$ in the norm L^2 (i.e., $\|g - g_N\|_{L^2} \to 0$ as $N \to \infty$). Let $f(x)$ and $f_N(x)$ be the corresponding solutions of the equation given before, i.e. $f(x) = Sg(x)$, $f_N(x) = Sg_N(x)$. Is it true that also $f_N(x)$ approximate $f(x)$ in the L^2 norm?

(4) If, in particular, $g_N(x)$ are obtained "truncating" the Fourier expansion of $g(x)$ (i.e., $g_N(x)$ is the Nth partial sum of the Fourier series defined at the beginning), show that $f_N(x)$ converges "rapidly" to the solution $f(x)$, meaning that there are constants $C_N \to 0$ as $N \to \infty$ such that

$$\|f - f_N\| \le C_N \|g - g_N\|$$

1.95

In the space $L^2(-\pi, \pi)$ consider the operator

$$T f = \frac{d^2 f}{dx^2} + f \quad \text{with periodic boundary conditions as in Problem 1.93}$$

(1) Find eigenvectors and eigenvalues of T with their degeneracy.

(2) Determine $\mathtt{Ker}\ T$.

(3) Consider the equation ($f(x)$ is the unknown and $g(x) \in L^2(-\pi, \pi)$ is given)

$$T f = g$$

Use expansions in Fourier series in terms of the eigenvectors found in (1). What condition must be imposed to the function $g(x)$ in order that some solution exists? For instance, does this equation admit solution if $g(x) = \cos^4 x$? and if $g(x) = \cos^3 x$? When the solution exists, is it unique? Write in the form of a Fourier series the most general solution.

(4) Considering more in general the operator

$$T_\alpha f = \frac{d^2 f}{dx^2} + \alpha f$$

for what $\alpha \in \mathbf{C}$ does T_α admit bounded inverse?

(5) Find $\|T_\alpha^{-1}\|$ if $\alpha = 1 + ia,\ a \in \mathbf{R}$.

1.96

In the space $H = L^2(0, 2\pi)$ consider the operator

$$T = \frac{d^2}{dx^2} \quad \text{with the boundary conditions } f'(0) = f'(2\pi) = 0$$

(1) Is T Hermitian (in a suitable domain)? Find eigenfunctions and eigenvalues of T. Show that the eigenfunctions provide an orthogonal complete system in H. Specify `Ker` T and `Ran` T.

(2) Using the Fourier expansion in terms of the eigenfunctions of T, specify for what $g(x) \in H$ the equation

$$T f = g$$

admits solution $f(x) \in H$. Is the solution unique? Given a $g(x)$ such that this equation admits solution, write as a Fourier series the most general solution $f(x)$.

(3) Is the solution obtained in (2) a continuous function? Is it possible to find a constant K such that, for any $x \in (0, 2\pi)$,

$$\left|\frac{df}{dx}\right| \le K\|g\|_{L^2} \ ?$$

1.97

In the space $H = L^2(0, \pi)$ consider the equation, where $f(x)$ is the unknown function and $\varphi(x)$ is given,

$$\frac{d^2 f}{dx^2} = \varphi(x) \quad \text{with vanishing boundary conditions}$$

(1) Write, using Fourier expansion in terms of the set $\{\sin nx,\ n = 1, 2, \ldots\}$, the (unique) solution $f(x)$. Find a constant C such that $\|f\| \le C\|\varphi\|$.

(2) Let $\varphi_\alpha(x)$ be a family of functions $\in H$ and let $f_\alpha(x)$ be the corresponding solutions; assume that $\varphi_\alpha(x) \to \psi(x)$ in the norm L^2, i.e. $\|\varphi_\alpha - \psi\|_{L^2} \to 0$ as $\alpha \to$

0, and let $h(x)$ be the solution corresponding to $\psi(x)$, i.e., $d^2h/dx^2 = \psi(x)$. It is true that $f_\alpha(x) \to h(x)$, and in what sense?

(3) Show that

$$\sup_{0\le x\le \pi} |f(x)| \le \big|(G, \Phi)\big|$$

where $G = G(x) \in H$ does not depend on $\varphi(x)$, and $\Phi(x) = \sum_{n\ge1} |\varphi_n| \sin nx$ where φ_n are defined by the Fourier expansion $\varphi(x) = \sum_n \varphi_n \sin nx$.

1.98

Consider the operator in $L^2(0, \pi)$

$$T = -\frac{d^2}{dx^2} + c\frac{d}{dx} \quad (c \in \mathbf{R}) \quad \text{with the boundary conditions } u(0) = u(\pi) = 0$$

(1) Find the eigenvalues and eigenvectors of T. Are the eigenvectors a complete set in $L^2(0, \pi)$?

(2) Does T admit a bounded inverse? If yes, calculate $\|T^{-1}\|$.

(3) Show that the eigenvectors are orthogonal with respect to the weight function $\rho = \exp(-cx)$, i.e., with respect to the scalar product defined by

$$(u, v)_\rho = \int_0^\pi \rho(x)u^*(x)v(x)dx \,.$$

(4) Verify that the problem is actually a classical Sturm–Liouville Problem

$$-\frac{1}{\rho}\left[\frac{d}{dx}\left(p(x)\frac{d}{dx}\right) + q(x)\right] = \lambda\, u$$

with $p = q = \cdots$: this confirms the orthogonality property seen before.

1.99

In the Hilbert space $L^2(1, \sqrt{3})$ consider the operator

$$T = x\frac{d}{dx} \quad \text{with the boundary condition } u(1) = u(\sqrt{3})$$

(1) Find the eigenvalues λ_n and eigenvectors $u_n(x)$ of T.

(2) Show that the eigenvectors $u_n(x)$ of T are not orthogonal, but that they turn out to be orthogonal with respect to the modified scalar product obtained introducing the weight function $\rho(x) = x$:

$$(f, g)_\rho = \int_1^{\sqrt{3}} x\, f^*(x)g(x)\, dx$$

Are the eigenvectors a complete set in H?

(3) Considering the expansion

$$f(x) = \sum_n c_n u_n(x)$$

give the formula for obtaining the coefficients c_n (use the results seen in (2)).

(4) Calculate $\sum_n |c_n|^2$ if in the above expansion $f(x) = x$.

(5) Does the expansion given in (3) provide a periodic function out of the interval $(1, \sqrt{3})$? If this expansion at the point, e.g., $x = 3/2$, converges to some value, at what points, out of the interval, does it take the same value?

1.100

(1) (*a*) In the space $L^2(0, +\infty)$ find, using integration by parts, the adjoint of the operator

$$A = x\frac{d}{dx}$$

(in a suitable dense domain of smooth functions chosen in such a way that the "finite part" $\Big[\ldots\Big]$ of the integral is zero).

(*b*) Show that the operator

$$\widetilde{A} = x\frac{d}{dx} + \frac{1}{2}I$$

is anti-Hermitian (i.e., $\widetilde{A}^+ = -\widetilde{A}$)

(2) Consider now the operator

$$T_\alpha f(x) = f\big((\exp\alpha)x\big) , \qquad \alpha \in \mathbf{R}$$

(*a*) Find $\|T_\alpha\|$.

(*b*) Show that one can find a real coefficient $c(\alpha)$ in such a way that the operator $\widetilde{T_\alpha} = c(\alpha)T_\alpha$ is unitary.

(3) Find the operators B and $\widetilde{B}$ defined by

$$\frac{d}{d\alpha}T_\alpha f\Big|_{\alpha=0} = Bf \ \text{ and } \ \frac{d}{d\alpha}\widetilde{T}_\alpha f\Big|_{\alpha=0} = \widetilde{B}f$$

compare with the operators A and $\widetilde{A}$ defined in (1) and interpret the results in terms of Lie groups and algebras of transformations (see also Problem 4.21).

1.101
Consider in the space $L^2(Q)$, where $Q = \{0 \le x \le \pi,\ 0 \le y \le \pi\}$ is the square of side π, the Laplace operator

$$T = \frac{\partial^2}{\partial x^2} + \frac{\partial^2}{\partial y^2}$$

with vanishing boundary conditions on the four sides of the square.

(1) Find eigenvalues and eigenvectors of T with their degeneracy.

(2) Show that the operator is invertible and find $\|T^{-1}\|$.

(3) Does the equation (where $g(x, y)$ is given and $f(x, y)$ the unknown)

$$Tf = g$$

admit solution for any $g(x, y) \in L^2(Q)$? Is the solution unique? (Use the Fourier expansion in terms of the eigenvectors of T.)

(4) If $g(x, y) = g(y, x)$, is the same property shared by the solution $f(x, y)$?

(5) Let S be the operator $Sf(x, y) = f(y, x)$. Find the common eigenvectors of T and S. Do they provide an orthogonal complete system in the space $L^2(Q)$?

1.102
Let Q be the square $\{0 \le x \le 2\pi,\ 0 \le y \le 2\pi\}$ and let T be the operator defined in $L^2(Q)$

$$T = \left(\frac{\partial}{\partial x} - \alpha\right)\left(\frac{\partial}{\partial y} - \beta\right), \qquad \alpha, \beta \in \mathbf{C}$$

with periodic boundary conditions:

$$u(x, 0) = u(x, 2\pi)\,; \qquad u(0, y) = u(2\pi, y)$$

(1) For what values of α, β is T Hermitian (in a suitable dense domain)?

(2) Using separation of the variables, find eigenvectors and eigenvalues of T (with arbitrary α, β); do the eigenvectors provide an orthogonal complete system for $L^2(Q)$?

(3) For what values of α, β is $\mathrm{Ker}\, T \neq \{0\}$?

(4) Fix now $\alpha = \beta = 0$. Using the results obtained in (2), specify for what $g(x, y) \in L^2(Q)$ the equation

$$Tf - f = g$$

admits solution $f(x, y) \in L^2(Q)$. Is the solution unique? Find the more general solution (if existing) with $g(x, y) = \sin 2(x + y)$.

1.103
Let Q be the square $\{0 \le x \le 2\pi,\ 0 \le y \le 2\pi\}$ and let T be the operator defined in $L^2(Q)$

$$T = \frac{\partial}{\partial x} + \alpha \frac{\partial}{\partial y}, \qquad \alpha \in \mathbf{R},\ \alpha \neq 0$$

with periodic boundary conditions (as in the previous problem)

(1) Using separation of the variables, find eigenvectors and eigenvalues of T; do the eigenvectors provide an orthogonal complete system for $L^2(Q)$?
(2) *(a)* Let $\alpha = 1$: find $\mathtt{Ker}\ T$, its dimension and find an orthogonal complete system for it.
(b) The same question if $\alpha = \sqrt{2}$.
(3) Using the results obtained in (2)*(a)*, write, as a Fourier expansion, the solution of the equation

$$\frac{\partial u}{\partial x} + \frac{\partial u}{\partial y} = 0$$

satisfying the condition

$$u(x, 0) = \begin{cases} 1 & \text{for } 0 < x < \pi \\ 0 & \text{for } \pi < x < 2\pi \end{cases}$$

Find in particular $u(x, \pi)$.
(4) Observing that the most general solution of the equation $u_x + u_y = 0$ is $u(x, y) = F(x - y)$ where F is arbitrary, it is easy to write the solution of the above equation satisfying the generic condition

$$u(x, 0) = \varphi(x) \in L^2(0, 2\pi)$$

1.2.6 Functionals

1.104
Study each one of the following functionals Φ: specify if the functional is bounded or not. In the case of bounded functionals, find their norm, the representative vector according to Riesz theorem, and the kernel. In the case of unbounded functionals, study their domain and kernel, which provide examples of *dense* subspaces in the Hilbert space. Therefore, these functionals are examples of not closed operators, see Problem 1.70, q. (4).

(1) In a Hilbert space with orthonormal complete system $\{e_n, n = 1, 2, \ldots\}$, and with $x = \sum_{n=1}^{\infty} a_n e_n$ consider

(a) $$\Phi(x) = \sum_{n=1}^{N} a_n \quad (N \geq 1) \quad \text{and} \quad \Phi(x) = \sum_{n=1}^{\infty} a_n$$

Recall Problems 1.16 and 1.62 to verify explicitly that the second functional is not a closed operator.
(b) for what sequences of complex numbers c_n, is the functional

$$\Phi(x) = \sum_{n=1}^{\infty} c_n a_n$$

a bounded functional?
(2) In $L^2(a,b)$ with $-\infty < a < b < \infty$ consider

$$\Phi(f) = \int_a^b f(x)\,dx \quad \text{and} \quad \Phi(f) = \int_a^b \exp(x) f(x)\,dx$$

(3) In $L^2(\mathbf{R})$ consider

(a) $$\Phi(f) = \int_{-\infty}^{+\infty} \exp(-|x|) f(x)\,dx$$

(b) $$\Phi(f) = \int_{-\infty}^{+\infty} f(x)\,dx \quad \text{and} \quad \Phi(f) = \int_{-\infty}^{+\infty} \sin x\; f(x)\,dx$$

(recall Problems 1.4 and 3.8 for what concerns the kernels)
(4) Consider, for different values of $\alpha, \beta, \gamma \in \mathbf{R}$,

$$\Phi(f) = \int_{-1}^{1} |x|^\alpha f(x)\,dx\;; \quad \int_1^{+\infty} x^\beta f(x)\,dx\;; \quad \int_{-\infty}^{+\infty} |x|^\gamma f(x)\,dx$$

For what values of α, β, γ are these functionals bounded?
(5) In $L^2(I)$, where I is any interval, consider

$$\Phi(f) = f(x_0)$$

which is defined in the subspaces of the functions continuous in a neighborhood of $x_0 \in I$. Let, e.g., $f(x) = 1\,, \forall x \in I$, then $\Phi(f) = 1$; construct a sequence of functions $f_n(x) \to f(x) = 1$ in the norm $L^2(I)$ such that $f_n(x_0) = 0$ (see Problem 1.6), or also a sequence $g_n(x) \to f(x) = 1$ in the norm $L^2(I)$ such that $g_n(x_0) \to +\infty$. Conclude: the functional is …

1.105
(1) In a Hilbert space H, with orthonormal set (not necessarily complete) $\{e_n\,, n = 1, 2, \ldots\}$, consider the functionals

$$\Phi_n(v) = a_n = (e_n, v)\,, \qquad v \in H$$

Study the convergence as $n \to +\infty$ of the sequence of functionals Φ_n.
(2) With the assumption that $\{e_n\}$ is complete, evaluate the sum of the series

$$S_v = \sum_n \Phi_n^*(w)\,\Phi_n(v)\,, \qquad v,\, w \in H$$

(3) Let now $H = L^2(-\pi, \pi)$ and let $f(x) \equiv f_1(x)$ be the function considered in Problem 1.20, q. (1), recalling its Fourier expansion obtained in that problem, find the sum of the series, where $g(x) \in L^2(-\pi, \pi)$ is given,

$$S_g = \sum_{m=0}^{+\infty} \frac{1}{2m+1} \Phi_{2m+1}(g)$$

1.106
Let $\Phi_a(f)$ be the functional defined in $L^2(0, 1)$

$$\Phi_a(f) = \frac{1}{\sqrt{a}} \int_0^a f(x)\,dx\,, \qquad 0 < a < 1$$

(1) Find $\|\Phi_a\|$.
(2) Show that $\Phi_a \to 0$ strongly (but not in norm) as $a \to 0$. *Hint*: restrict first to the dense subspace of continuous functions, then ....

1.107
Let $\Phi_a(f)$ be the functional defined in $L^2(0, +\infty)$

$$\Phi_a(f) = \frac{1}{\sqrt{a}} \int_0^a f(x)\,dx\,, \qquad a > 0$$

(1) Find $\|\Phi_a\|$.
(2) Show that $\Phi_a \to 0$ strongly (but not in norm) as $a \to +\infty$. *Hint*: restrict first to the dense subspace of functions with compact support, then ....

1.2.7 Time-Evolution Problems. Heat Equation

1.108
Consider the heat equation, also called diffusion equation, in $L^2(0, \pi)$

$$\frac{\partial u}{\partial t} = \frac{\partial^2 u}{\partial x^2}\,, \qquad u = u(x, t)$$

with vanishing boundary conditions

$$u(0, t) = u(\pi, t) = 0$$

and the (generic) initial datum

$$u(x,0) = f(x) \in L^2(0,\pi)$$

(1) Find the time-evolution $u = u(x,t)$ for $t > 0$. *Hint*: write $f(x)$ and $u(x,t)$ as Fourier expansions in terms of the orthogonal complete system $\{\sin nx,\ n = 1, 2, \ldots\}$.

(2) Show that the solution $u(x,t)$ tends to zero (in the $L^2(0,\pi)$ norm) as $t \to +\infty$. How "rapidly" does it tend to zero?

(3) Show that for any $t > 0$ the solution $u(x,t)$ is infinitely differentiable with respect to x and to t.

1.109

Considering the time-evolution problem proposed above, let E_t be the "time-evolution operator" defined by

$$E_t\ :\ u(x,0) \to u(x,t)\,, \qquad t > 0$$

(1) Find eigenvalues and eigenvectors of E_t.

(2) Calculate $\|E_t\|$ and verify that $\|E_t\| \to 0$ as $t \to +\infty$.

(3) Show that $E_t \to I$ in the strong sense, not in norm, as $t \to 0^+$.

1.110

Consider the heat equation in $L^2(-\pi,\pi)$, now with periodic boundary conditions

$$u(-\pi,t) = u(\pi,t), \quad u_x(-\pi,t) = u_x(-\pi,t)$$

and initial condition

$$u(x,0) = f(x) \in L^2(-\pi,\pi)$$

(1) Find the time-evolution $u(x,t)$: use now Fourier expansions with respect to the orthogonal complete system $\{\exp(inx),\ n \in \mathbf{Z}\}$.

(2) What happens as $t \to +\infty$?

(3) If $\int_{-\pi}^{\pi} u(x,0)\,dx = 0$, is this property preserved for all $t > 0$, i.e. $\int_{-\pi}^{\pi} u(x,t)$ $dx = 0, \forall t > 0$?

(4) Study the convergence as $t \to +\infty$ and as $t \to 0^+$ of the operator E_t defined in the previous problem.

1.111

Consider the case of a non-homogeneous equation in $L^2(-\pi,\pi)$ of the form

$$\frac{\partial u}{\partial t} = \frac{\partial^2 u}{\partial x^2} + F(x)$$

with periodic boundary conditions as in previous problem, and initial datum

$$u(x,0) = f(x)$$

where $F(x)$ and $f(x) \in L^2(-\pi, \pi)$. Find (in the form of a Fourier expansion) the time-evolution $u(x,t)$. *Hint*: write $F(x)$ and $f(x)$ in the form of a Fourier expansion: $F(x) = \sum_{n\in\mathbf{Z}} F_n \exp(inx)$, $f(x) = \sum_{n\in\mathbf{Z}} f_n \exp(inx)$ and look for the solution writing $u(x,t) = \sum_{n\in\mathbf{Z}} a_n(t) \exp(inx)$, then deduce a differential equation for $a_n(t)$.

1.112

(1) Consider an orthonormal complete system $\{e_n,\ n = 1, 2, \ldots\}$ in a Hilbert space H and let T be the linear operator defined by

$$T\, e_n = -n^2 e_n$$

Find the time-evolution of the problem for $v = v(t) \in H$

$$\frac{d}{dt} v = T\, v$$

with the generic initial condition $v(0) = v_0 \in H$.

(2) The same problem with $n \in \mathbf{Z}$.

(3) Study the properties of the time-evolution operator E_t defined by $E_t : v_0 \to v(t)$. (This is just the "abstract" version of the Problems 1.108–1.110.)

1.113

Consider the time-evolution problem, in a Hilbert space H where $\{e_n,\ n \in \mathbf{Z}\}$ is an orthonormal complete system

$$\frac{d}{dt} v = T v\,, \qquad v = v(t) \in H$$

with the linear operator T defined by

$$T\, e_0 = e_0 \ \ ; \ \ T\, e_n = e_{-n}$$

(1) Find the time-evolution if the initial condition is given by $v(0) = e_0$

(2) The same if $v(0) = e_1$

(3) Extend to the case where $v(0)$ is a generic vector $v \in H$.

1.114

The same problem as before with

$$T\, e_n = n\, e_{-n}$$

(1) Find the time-evolution if the initial condition is given by $v(0) = e_0$

(2) The same if $v(0) = e_n$

(3) Extend to the case where $v(0)$ is a generic vector $v \in H$.

1.115

In the space $L^2(0, 2\pi)$ consider the operator

$$T = \frac{d}{dx} \quad \text{with the boundary condition} \quad f(2\pi) = -f(0)$$

(1) Find the eigenvectors $u_n(x)$ and the eigenvalues of T. Are the eigenvectors orthogonal? Show that the eigenvectors are a complete set in $L^2(0, 2\pi)$.

(2) Using the series expansion $\sum_n a_n(t)u_n(x)$, write in the form of series the solution of the equation

$$\frac{d}{dt} f = T f \quad \text{with the generic initial condition} \quad f(x, 0) = f_0(x) \in L^2(0, 2\pi)$$

(3) Show that this solution is periodic in time: what is the period?

(4) Let E_t be the "time-evolution" operator

$$E_t \,:\, f(x, 0) \to f(x, t)$$

verify that the eigenvectors of T are also eigenvectors of E_t. Is E_t unitary?

(5) Study the convergence of the operators E_t to the identity operator I as $t \to 0$.

1.116

(1) Let A be the 2×2 matrix

$$A = \begin{pmatrix} 1 & 0 \\ 1 & 1 \end{pmatrix}$$

Evaluate $\exp(At)$ and solve the time-evolution problem

$$\dot{u} = Au\,, \qquad u = u(t) \in \mathbf{R}^2$$

with the generic initial condition $u(0) = a \in \mathbf{R}^2$.

(2) Generalize: consider a Hilbert space H with orthonormal complete system $\{e_n,\ n = 1, 2, \ldots\}$ and the linear operator B defined by

$$B\, e_n = \begin{cases} e_n + e_{n+1} & \text{for} \quad n = 1, 3, 5, \ldots \\ e_n \quad \text{for} \quad n = 2, 4, \ldots \end{cases}$$

(*a*) evaluate $\exp(Bt)$. *Hint*: it can be useful to write $B = I + \widetilde{B}$, then $\widetilde{B}^2 = \cdots$.

(*b*) find the solution of the time-evolution problem, where $u = u(t) \in H$,

$$\dot{u} = Bu$$

if $u(0) = u_0 = e_1$ and if $u_0 = e_2 + e_3$
(c) write the solution with a generic initial condition $u(0) = u_0 \in H$.

1.117
Consider the heat equation in $L^2(0, \pi)$ with "wrong" sign:

$$\frac{\partial^2 u}{\partial x^2} = -\frac{\partial u}{\partial t}$$

(or, equivalently, the heat equation after time-inversion) with vanishing boundary conditions $u(0, t) = u(\pi, t) = 0$. Let $f(x) \in L^2(0, \pi)$ denote the initial condition: $f(x) = u(x, 0)$.

(1) Are there initial conditions $f(x)$ such that the L^2-norm of the corresponding solution $u(x, t)$ remain bounded for all $t > 0$? (i.e., is there a constant C such that $\|u(x, t)\| < C,\ \forall t > 0$?)

(2) Show that there is a dense set of initial conditions $f(x)$ such that for any fixed $t > 0$ the solution exists (i.e., $u(x, t) \in L^2(0, \pi)$ for any fixed $t > 0$).

(3) Give an example of a continuous $f(x)$ such that the solution does not exist (i.e., $\notin L^2(0, \pi)$), $\forall t > 0$.

(4) Let $f_N(x)$ be a sequence of functions converging to zero in the L^2-norm, i.e., $\|f_N(x)\|_{L^2(0,\pi)} \to 0$ as $N \to \infty$. Does this imply that the same is true for the corresponding solutions $u_N(x, t)$ for any fixed $t > 0$?

1.118
Consider the d'Alembert equation, also called wave equation,

$$\frac{\partial^2 u}{\partial x^2} = \frac{\partial^2 u}{\partial t^2}, \qquad u = u(x, t),\ 0 \le x \le \pi,\ t \in \mathbf{R}$$

describing, e.g., the displacements of a vibrating string. Assume vanishing boundary conditions

$$u(0, t) = u(\pi, t) = 0$$

and initial data of the form

$$u(x, 0) = f(x) \in L^2(0, \pi) \quad \text{and} \quad u_t(x, 0) = 0$$

(1) Show that the solution can be written in the form

$$u(x, t) = \sum_{n=1}^{\infty} a_n \sin nx \cos nt \quad \text{where} \quad a_n = \cdots$$

(2) Let now

$$u(x,0) = f(x) = \begin{cases} \sin 2x & \text{for } 0 \le x \le \pi/2 \\ 0 & \text{for } \pi/2 \le x \le \pi \end{cases}$$

Expanding $f(x)$ in Fourier series in terms of the orthogonal complete system $\{\sin nx,\ n = 1, 2, \ldots\}$, show that the Fourier expansion of the solution has the form

$$u(x,t) = (1/2)\sin 2x\ \cos 2t + u_1(x,t)$$

where $u_1(x,t) = \sum_n \ldots$ and where this series contains only terms with *odd* n. Comparing with the graph of $u(x,0)$, it is easy to deduce the graph of $u_1(x,0)$. Show that $u_1(x,\pi) = -u_1(x,0)$ and deduce $u(x,\pi)$. Comparing then the graphs of $u(x,0)$ and of $u(x,\pi)$ confirm a well-known property of waves propagating in an elastic string with vanishing boundary conditions.

1.2.8 Miscellaneous Problems

1.119
In the Hilbert space $L^2(\mathbf{R})$ consider the operator

$$T_a f(x) = \begin{cases} -f(x) & \text{for } x < a \\ f(x) & \text{for } x > a \end{cases}, \qquad a \in \mathbf{R}$$

(1) Find the eigenvalues and eigenvectors of T_a.

(2) Study the convergence of the family of operators T_a as $a \to \infty$

(3) Let $a, b \in \mathbf{R}$ with $a \neq b$. Is it possible to write the operator $T_a T_b$ as a combination of projections?

(4) Is it possible to have an orthogonal complete system of simultaneous eigenvectors of T_a and T_b?

1.120
In the space $H = L^2(0, 2\pi)$ let $v_n = \exp(inx), n \in \mathbf{Z}$, and let T be the operator

$$T\ f(x) = \frac{1}{\pi}\int_0^\pi f(x+y)\,dy$$

where the functions $f(x)$ are periodically prolonged with period 2π.

(1) Find $T(v_n)$, and then the eigenvectors and eigenvalues (with their degeneracy) of T (do not forget the case $n = 0$!)

(2) Show that $T(f)$ is a continuous function, for any $f \in H$.

(3) Is T compact?

(4) Find $\|T^N\|$ where N is any integer.

(5) Study the convergence of the sequence of operators T^N as $N \to \infty$.

1.121

Let T be the operator in $L^2(\mathbf{R})$

$$T\,f(x) = \alpha f(x) + \beta f(-x)\,, \qquad \alpha, \beta \in \mathbf{C}\,;\ \alpha, \beta \neq 0$$

(1) Find $\|T\|$

(2) Find eigenvalues and eigenvectors of T. Is there an orthogonal complete system of eigenvectors of T?

(3) Under what conditions on α, β is T unitary? Let e.g. $\alpha = 1/\sqrt{2}$: find β in order to have T unitary.

(4) Let $\alpha = 2/3$, $\beta = 1/3$: study the convergence as $n \to \infty$ of the sequence of operators T^n.

1.122

In the Hilbert space $H = L^2(0, +\infty)$ consider the two operators

$$T_n = \frac{x}{1+x^2}\,P_n \quad \text{and} \quad S_n = \sin \pi x\, P_n\,, \qquad n = 1, 2, \ldots$$

where P_n is the projection on the Hilbert subspace $L^2(0, n)$.

(1) Specify if T_n and S_n are Hermitian. Find $\|T_n\|$ and $\|S_n\|$.

(2) *(a)* Fixed n, for what $g(x) \in H$ does the equation (with $f(x) \in H$ the unknown)

$$T_n\, f = g$$

admit solution? Is the solution (when it exists) unique?

(b) Is $\mathtt{Ran}\, T_n$ a Hilbert subspace of H? Is it true that $H = \mathtt{Ker}\, T_n \oplus \mathtt{Ran}\, T_n$?

(3) The same question as in (2) *(a)* for the equation

$$T_n\, f = S_n\, g$$

(4) Study the convergence as $n \to \infty$ of the two sequences of operators T_n and S_n.

1.123

In the space $H = L^2(0, 2\pi)$, with the functions periodically prolonged with period 2π out of this interval, let T be the operator

$$T\,f(x) = f(x - \pi/2)$$

(1) Find T^4. What information can be deduced about the eigenvalues of T?

(2) Show that $e_n = \exp(inx),\ n \in \mathbf{Z}$ are eigenvectors of T, find the eigenvalues with their degeneracy. Is $|\sin 2x|$ an eigenfunction of T?

(3) Let $S = T + T^2$; show that $S = T^{-2}(I + T^{-1})$. Find $\mathtt{Ker}\, S$ and $\mathtt{Ran}\, S$ (expand in terms of the eigenvectors obtained in (2)). Is it true that $\mathtt{Ran}\, S$ is a Hilbert subspace of H and that $H = \mathtt{Ran}\, S \oplus \mathtt{Ker}\, S$?

1.124

Let T be the operator defined in the space $L^2(-\pi, \pi)$ (the functions must be periodically prolonged with period 2π out of the interval $(-\pi, \pi)$)

$$T\,f(x) = \frac{1}{2i}\left(f\left(x + \frac{\pi}{2}\right) - f\left(x - \frac{\pi}{2}\right)\right)$$

(1) Show that T is a combination of two unitary operators.

(2) Find eigenvectors and eigenvalues (with their degeneracy) of T (use the complete system $\{e_n = \exp(inx),\ n \in \mathbf{Z}\}$).

(3) Show that T is a combination of two projections (on what subspaces?)

(4) Find $\|T\|$, $\|T + (1 + 2i)I\|$ and $\|(T + (1 + 2i)I)^{-1}\|$

1.125

In the space $H = L^2(-1, 1)$ let T be the operator

$$T\,f(x) = \int_{-1}^{1} K(x, y)\, f(y)\, dy \quad \text{where} \quad K(x, y) = 1 + xy$$

(1) Find $\mathtt{Ran}\, T$ with its dimension, and find an orthonormal complete system for it.

(2) Replace now (in this question) K with $K_{AB} = A + B\,xy$ where A, B are constants. Is it possible to choose $A, B \neq 0$ in such a way that the corresponding operator T_{AB} becomes a projection? *Hint*: a projection behaves as the identity on its range ...

(3) Find eigenvalues and eigenvectors of T, with their degeneracy.

(4) Find $\|T\|$.

(5) The same questions (1) and (2) with $K = 1 + x^2y^2$ and with $K_{AB} = A + B\,x^2y^2$.

1.126

Consider the operator defined in $L^2(-\pi, \pi)$

$$T\,f(x) = \alpha f(x) + \beta \int_{-\pi}^{\pi} f(y)\, dy\,, \qquad \alpha, \beta \in \mathbf{C};\ \alpha, \beta \neq 0$$

(1) *(a)* Find eigenvalues and eigenvectors of T, with their degeneracy, and specify if the eigenvectors provide an orthogonal complete system for $L^2(-\pi, \pi)$. *(b)* Find $\|T\|$

(2) Let now $\alpha = 1$: is it possible to choose β in such a way that T is a projection?

(3) Study the convergence as $n \to \infty$ of the sequence of operators T_n defined by

$$T_n f(x) = \alpha f(x) + \beta \exp(inx) \int_{-\pi}^{\pi} f(y)\, dy$$

1.127

Consider in the space $H = L^2(-\pi, \pi)$ the operator

$$T_a f(x) = \frac{1}{2a} \int_{x-a}^{x+a} f(y)\, dy = \frac{1}{2a} \int_{-a}^{a} f(x+y)\, dy\,, \qquad 0 < a \le \pi$$

where the function $f(x)$ must be periodically prolonged with period 2π.

(1) Let D be the operator

$$D = \frac{d}{dx} \quad \text{with periodic boundary condition } f(-\pi) = f(\pi)$$

(a) Find the eigenvalues and the eigenfunctions of D, with their degeneracy.
(b) Show that $T_a D = D T_a$ (in a dense domain).

(2) Using the results obtained in (1), find the eigenvalues and eigenvectors of T_a, with their degeneracy.

(3) Using the results obtained in (2), specify if T_a is Hermitian and if it is compact.

(4) Expanding $f(x)$ as Fourier series of the eigenfunctions of T_a, write $T_a f(x)$ for a generic $f(x) \in H$, and:

(a) Let $a = \pi$: find $T_\pi f$ and $\mathtt{Ran}\, T_\pi$.

(b) Let $a = \pi/2$: for what (integer) values of m does the function $g(x) = 1 + \sin mx$ belong to $\mathtt{Ran}\, T_a$?

(c) Let $a = 1$: is $\mathtt{Ran}\, T_a$ a dense subspace in H? does it coincide with H? is T_a invertible?

(5) Find $\|T_a\|$. Study the convergence as $a \to 0$ of the family of operators T_a.

1.128

Consider the operator T_a defined in $L^2(0, 2\pi)$, with the functions periodically prolonged with period 2π out of this interval,

$$T_a f(x) = f(x+a) - f(x)\,, \qquad 0 < a \le 2\pi$$

and the operator (see previous problem, q.(1))

$$D = \frac{d}{dx} \quad \text{with periodic boundary condition } f(0) = f(2\pi)$$

(1) Find eigenfunctions and eigenvalues (with their degeneracy) of the operator D.

(2) Observing that $DT_a = T_a D$ (in a dense domain) find eigenvectors and eigenvalues of T_a. What is the degeneracy of the eigenvalues if $a = \pi/2$? and if $a = 1$?

(3) Show that T_a is a normal operator for any a. Find $\|T_a\|$.

(4) For what choice of a does the operator $T_a - iI$ admit bounded inverse? Find $\|(T_a - iI)^{-1}\|$ if $a = \pi/2$ and if $a = 1$.

1.129

Consider the operator defined in $H = L^2(0, +\infty)$

$$T\,f(x) = \frac{1}{x} f\left(\frac{1}{x}\right)$$

(1) Find $\|T\|$ and T^{-1}.

(2) Is T unitary?

(3) Observing that $T^2 = \cdots$, it is easy to find the eigenvalues and construct the eigenfunctions of T .... Compare with the next problem.

1.130

Let now T be the operator defined in $H = L^2(0, +\infty)$

$$T\,f(x) = f\left(\frac{1}{x}\right)$$

(1) Is T bounded?

(2) Specify what among the following functions $\in H$ belong to the domain of T:

$$f_1(x) = \begin{cases} 1 & \text{for } 0 < x < 1 \\ 0 & \text{for } x > 1 \end{cases} \quad ; \quad f_2(x) = \begin{cases} x & \text{for } 0 < x < 1 \\ 0 & \text{for } x > 1 \end{cases} \quad ;$$

$$f_3(x) = \exp(-x) \quad ; \quad f_4(x) = \begin{cases} x & \text{for } 0 < x < 1 \\ 1/(1+x) & \text{for } x > 1 \end{cases}$$

(3) As in the case of the operator considered in the problem above, one has $T^2 = I$. But in the present case it is not obvious to find its eigenfunctions, indeed only when $f(x)$ belongs to its domain, one can apply the usual argument: for any $f(x)$, then $f(x) \pm Tf(x)$ are eigenfunctions. Find explicitly at least some eigenfunction of T.

1.131

(1) Let T be the operator defined in $L^2(0, 2\pi)$

$$T\,f(x) = \int_0^x f(t)\,dt\,, \qquad 0 < x \le 2\pi$$

(a) find the matrix elements $T_{nm} = (e_n, Te_m)$ of T with respect to the orthonormal complete system $\{e_n(x) = \exp(inx)/\sqrt{2\pi},\ n \in Z\}$.
(b) is T bounded?
(c) look for the eigenvalues of T (differentiate first both members of the equation $Tf(x) = \lambda f(x)$)
(2) The same questions $(b), (c)$ for the operator defined in $L^2(-\infty, 1)$

$$T\, f(x) = \int_{-\infty}^{x} f(t)\, dt\,, \qquad -\infty < x < 1$$

(3) The same questions $(b), (c)$ for the operator given in (2) but defined in $L^2(\mathbf{R})$.

1.132
In the space $H = L^2(Q)$, where Q is the square $0 \le x \le 1,\ 0 \le y \le 1$, introduce the orthonormal complete set $\{e_{nm} = \exp(2in\pi x)\exp(2im\pi y),\ n, m \in \mathbf{Z}\}$, which can be more conveniently written, now with $n, m \ne 0$:

$$e_{00} = 1;\ e^{(1)}_{n0} = \exp(2in\pi x);\ e^{(2)}_{0m} = \exp(2im\pi y);\ \widetilde{e}_{nm} = \exp(2in\pi x)\exp(2im\pi y)$$

(1) Accordingly, with clear notation, show that H can be decomposed as a direct sum

$$H = H^{(0)} \oplus H^{(1)} \oplus H^{(2)} \oplus \widetilde{H}$$

where $H^{(0)}$ is 1-dimensional, etc. For any $f^{(1)}(x) \in H^{(1)},\ \widetilde{f}(x, y) \in \widetilde{H}$, evaluate

$$\int_0^1 f^{(1)}(x)\, dy;\ \int_0^1 f^{(1)}(x)\, dx;\ \int_0^1 \widetilde{f}(x, y)\, dy$$

(2) Let T be the operator defined in H

$$T\, f(x, y) = \int_0^1 f(x, y)\, dy$$

Evaluate T^2. Find $\mathtt{Ran}\, T$, $\mathtt{Ker}\, T$ and $\|T\|$.
(3) For fixed nonzero $k \in \mathbf{Z}$, let T_k be defined by

$$T_k\, f(x, y) = \int_0^1 f(x, y)\exp(-2ik\pi y)\, dy$$

Evaluate T_k^2. Find $\mathtt{Ran}\, T_k$, $\mathtt{Ker}\, T_k$.
(4) Study the convergence as $k \to +\infty$ of the sequence of operators T_k

Chapter 2
Functions of a Complex Variable

2.1 Basic Properties of Analytic Functions

2.1
(1) Evaluate

$$(1+i)^{100}\,; \quad \sqrt[3]{\sqrt{1-i}}\,; \quad \left(\frac{1}{1+i\sqrt{3}}\right)^{21/2}$$

(2) Solve the equations, with $z \in \mathbf{C}$,

$$\sin z = 4i\,; \quad \cos z = 3\,; \quad \exp z = \pm 1/e\,; \quad \cosh z := \frac{\exp z + \exp(-z)}{2} = -1$$

2.2
The following functions of the complex variable $z \in \mathbf{C}$

$$z^2 \sin(1/z)\,; \quad \exp(-1/z^4)\,; \quad z\exp(-1/z^2)\,; \quad z\exp(i/z)$$

have, as well-known, an essential singularity at $z = 0$, therefore their limit as $z \to 0$ does not exist. Verify explicitly that this limit does not exist showing that these functions assume different values approaching arbitrarily near $z = 0$ along different paths, or sequences of points.

2.3
(1) Write the most general function $f(z)$ which has a pole of order 2 at $z = i$ with residue $-3i$ and is analytic in all other points, including the point $z = \infty$. What changes without the assumption of analyticity at $z = \infty$?

G. Cicogna, *Exercises and Problems in Mathematical Methods of Physics*, Undergraduate Lecture Notes in Physics,
https://doi.org/10.1007/978-3-030-59472-5_2

(2) A function $f(z)$ is analytic $\forall z \in \mathbf{C}$ apart from the point $z = \infty$ and the point $z = 1$ where the residue is 1 and where

$$\lim_{z\to 1}(z-1)^3 f(z) = 2$$

(i) Consider the second derivative $f''(z)$: determine what is its singularity and its residue at $z = 1$. Are there any other singularities (apart from $z = \infty$)? And can $f''(z)$ be analytic at $z = \infty$?

(ii) What about the singularities of the primitive $F(z)$ of $f(z)$ (i.e.: $F'(z) = f(z)$)?

2.4

Let $f(z)$ be an analytic function for all complex z (apart from $z = \infty$).

(1) It is known that its real part $u(x, y) = \mathrm{Re}\, f$ has the form

$$u(x, y) = a(x) + b(y)$$

What is the most general $f(z)$ satisfying this condition?

(2) The same question if

$$u(x, y) = a(x)\, b(y)$$

2.5

Expand the function

$$f(z) = \frac{1}{1-z}$$

in Taylor-Laurent power series, and specify the respective regions of convergence

(a) in a neighborhood of $z_0 = 0$; *(b)* in a neighborhood of $z_0 = 2i$;

(c) in a neighborhood of $z_0 = \infty$; *(d)* in a neighborhood of $z_0 = 1$

2.6

(1) Find the sum of the series

$$\sum_{n=1}^{\infty} n z^n$$

(2) Determine the radius R of convergence of the series, where $p \in \mathbf{R}$ is fixed

$$\sum_{n=1}^{\infty} n^p z^n$$

(3) Determine the region of convergence of the series

$$\sum_{n=0}^{\infty}\left(\frac{z}{3}\right)^n+\sum_{n=0}^{\infty}\left(\frac{2}{z}\right)^n$$

2.7
For what $z \in \mathbf{C}$ is the series

$$\sum_{n=0}^{\infty}\exp(-nz)$$

convergent? Find its sum $S(z)$ and determine the singularities (including the point $z=\infty$) of the function $S(z)$ extended to the whole complex plane $\mathbf{C}$.

2.8
Determine the singularity at point $z=\infty$ and its residue at point $z=\infty$ of the function

$$f(z)=\frac{z\sin z}{a_4z^4+a_2z^2+a_0}\,, \qquad a_0, a_2, a_4\neq 0$$

Hint: $f(z)$ is an *even* function of z, then it contains only powers of $z^2, \ldots$.

2.9
Verify the validity of the l'Hôpital theorem in its simplest form in the case:

$$\lim_{z\to z_0}\frac{f(z)}{g(z)}$$

where $f(z)$ and $g(z)$ are analytic in a neighborhood of z_0 and $f(z_0)=g(z_0)=0$.

2.10
Determine the singularities of the function

$$f(z)=\frac{z^2+\pi^2}{1+\exp z}$$

and specify the radius R of convergence of the Taylor expansion of $f(z)$ around the origin $z_0=0$.

2.11
Determine the singularities in the complex plane z (included the point $z=\infty$) of the functions

$$f_1(z)=\frac{1}{\sin z}\quad ; \quad f_2(z)=\frac{1}{\sin(1/z)}$$

2.12

(1) Find the coefficients $a_{-1},\ a_0,\ a_1$ and a_2 of the Taylor-Laurent expansion $\sum_n a_n z^n$ in the neighborhood of the origin $z_0 = 0$ of the function

$$f(z) = \frac{1}{\sin z}$$

Hint: find first a_{-1}, then the function $f(z) - (a_{-1}/z)$ is analytic around $z_0 = 0$.

(2) Show that the function

$$f(z) = \frac{1}{1 - \cos z} - \frac{2}{z^2}$$

is analytic in a neighborhood of $z_0 = 0$ and find the coefficients a_0 and a_1 of its Taylor expansion.

2.13

Determine the singularities in the complex plane z (included the point $z = \infty$) of the functions

$$f_1(z) = \frac{\sin^2 z}{z^4 (z - \pi)(z + 2\pi)^2}\ ; \quad f_2(z) = \frac{\sin z}{z^2} - \frac{\cos z}{z} - \frac{z}{3}$$

2.14

Determine the singularities in the complex plane z (included the point $z = \infty$) and the region of convergence of the Taylor-Laurent expansion around $z_0 = 1$ of the function

$$f(z) = \frac{\exp(z^2) + \exp(-z^2) - 2 - z^4}{z^n (z - 1)}$$

depending on the values of the integer number n.

2.15

Determine the singularities in the complex plane (including the point $z = \infty$) of the following functions

$$f_1(z) = \sin\sqrt{z}\ ; \quad f_2(z) = \sin^2\sqrt{z}\ ; \quad f_3(z) = \cos\sqrt{z}\ ;$$

$$f_4(z) = \frac{\sin\sqrt{z}}{z}\ ; \quad f_5(z) = \frac{\sin\sqrt{z}}{\sqrt{z}}\ ; \quad f_6(z) = \frac{\cos\sqrt{z}}{\sqrt{z}}\ ;$$

$$f_7 = \frac{\sin\sqrt{z}}{z^2} - \frac{\cos\sqrt{z}}{z}\ ; \quad f_8(z) = z \sin\frac{1}{\sqrt{z}}\ ; \quad f_9(z) = \sqrt{z}\,\sin\frac{1}{\sqrt{z}}$$

2.16
Find the branch points of the functions

$$f_1(z) = \log z - \log(z-1)\,; \quad f_2(z) = \log z + \log(z-1)\,;$$

$$f_3(z) = \sqrt{z(z-1)}; \quad f_4(z) = \sqrt[3]{z(z-1)}; \quad f_5(z) = \sqrt[3]{z^2(z-1)}$$

2.17
(1) For what positive integer n does the function

$$f_1(z) = \frac{\sin z - z + \sin(z^3/6)}{z^n(1-z^2)^2}$$

admit Taylor expansion (with positive powers of z) in the neighborhood of $z = 0$? What is the behavior of the function at $z = \infty$?

(2) What changes for the function

$$f_2(z) = \frac{\sin z - z + \sin(z^3/6)}{z^n\sqrt{1-z^2}} \; ?$$

2.18
(1) For what values of $\alpha \in \mathbf{C}$ (if any, $\alpha \neq 0$) is the function

$$f_1(z) = \frac{\exp(\alpha z) - \exp(-\alpha z)}{1-z^2}$$

analytic for all $z \in \mathbf{C}$ (apart from $z = \infty$)? What happens at $z = \infty$?

(2) The same question for the function

$$f_2(z) = \frac{\exp(\alpha z) - \exp(-\alpha z)}{\sqrt{1-z^2}}$$

2.19
Fix the cut line of the functions $\log z$ and $\sqrt{z}$ along the positive real axis.

(1) Check if the following identity is true

$$\log(z^3) = 3\log z$$

(2) Determine the other singularities (besides the cut line) of the functions

$$f_1(z) = \frac{1}{\sqrt{z}-i}\,; \quad f_2(z) = \frac{1}{\sqrt{z}+i}\,; \quad f_3(z) = \frac{i\pi + \log z}{z+1}\,; \quad f_4(z) = \frac{i\pi - \log z}{z+1}$$

2.20

Despite the presence of branch points of the following functions, show that it is possible to choose conveniently the cut lines in such a way to have functions analytic at $z = 0$. Evaluate the first terms of their Taylor expansions:

$$f_1(z) = \sqrt{1 \pm z^2}\,; \quad f_2(z) = \log\frac{1+z}{1-z}\,; \quad f_3(z) = \log(1 \pm z^2)$$

2.2 Evaluation of Integrals by Complex Variable Methods

2.21

(1) Show that if a function $f(z)$ has at the point $z = \infty$ a zero of order ≥ 2 then its residue $\mathrm{R}(\infty)$ is zero.

(2) As well-known, the integral of a rational function

$$\int_{-\infty}^{+\infty} \frac{P(x)}{Q(x)}\, dx$$

where $P(x)$, $Q(x)$ are polynomials of the real variable x, exists if *(i)* $Q(x)$ has no (real) zeroes and *(ii)* the degrees n_P, n_Q of $P(x)$, $Q(x)$ satisfy the condition $n_Q \geq n_P + 2$. Show that this second condition implies that the residue $\mathrm{R}(\infty)$ at the point $z = \infty$ of the complex function $f(z) = P(z)/Q(z)$ is zero.

(3) To evaluate the above integral with the method of residues, one can consider a closed contour consisting of a segment $-R \leq x \leq R$ along the real axis and a semicircle of radius R, either in the *upper* or in the *lower* complex plane $\mathbf{C}$ (and then let $R \to \infty$). Show that the property $\mathrm{R}(\infty) = 0$ ensures that—as expected!—the result of the integration does not depend on the choice about the closing contour.

2.22

Evaluate the integrals

$$\int_{-\infty}^{+\infty} \frac{x^2}{(x^2+1)^2}\, dx\,; \qquad \int_{-\infty}^{+\infty} \frac{x}{(x+i)(x-2i)^2(x-3i)}\, dx$$

2.23

Evaluate the integrals

$$\oint_{|z|=2} \frac{\sin z}{(z-1)^5}\, dz\,; \qquad \oint_{|z-2|=3} \frac{z}{\sin^2 z}\, dz\,; \qquad \oint_{|z|=5} \frac{1-\exp z}{1+\exp z}\, dz$$

2.24
Evaluate the integrals

$$\oint_{|z|=1} \frac{z^2}{(2z-1)(z^2+2)}\,dz\,; \qquad \oint_{|z|=1} \frac{\exp z}{z}\,dz$$

Integrals of this type, which can be very easily evaluated in the complex plane, produce nontrivial results when z is replaced by $\exp(i\theta)$: verify!

2.25
Evaluate the integrals (put $\exp(i\theta) = z$ and transform the integrals into integrals along the circle $|z| = 1$ in the complex plane):

$$\int_0^{2\pi} \frac{\cos\theta}{2+\cos\theta}\,d\theta\,; \quad \int_0^{2\pi} \frac{1}{1+\cos^2\theta}\,d\theta\,; \quad \int_0^{2\pi} \frac{\sin\theta}{(2+\sin\theta)^2}\,d\theta$$

2.26
Evaluate the integrals, using Jordan lemma,

$$\int_{-\infty}^{+\infty} \frac{\exp(\pm iax)}{(x-i)^2}\,dx \quad (a>0)\,; \qquad \int_{-\infty}^{+\infty} \frac{\exp(ix)}{(x^2+1)^2}\,dx$$

2.27
Evaluate the following integrals; here, the functions $\sin x$, $\cos x$ must be replaced by $\exp(\pm ix)$, in order that the semicircular closing contour of integration in the complex plane gives a vanishing contribution, according to Jordan lemma:

$$\int_{-\infty}^{+\infty} \frac{\cos x}{1+x^2}\,dx\,; \quad \int_{-\infty}^{+\infty} \frac{\sin x}{1+x+x^2}\,dx$$

2.28
Evaluate the following integrals; in these cases, the resort to Jordan lemma produces the appearance of singularities (simple poles) along the real axis, and therefore the necessity of introducing one or more "indentations" along the real axis: see Fig. 2.1, which refers to the first one of the following integrals

$$\int_{-\infty}^{+\infty} \frac{\sin x}{x(1+x^2)}\,dx\,; \qquad \int_{-\infty}^{+\infty} \frac{\sin x}{x}\,dx\,;$$

$$\int_{-\infty}^{+\infty} \frac{\cos(\pi x/2)}{(1-x)(x^2+1)}\,dx\,; \qquad \int_{-\infty}^{+\infty} \frac{\sin(\pi x)}{x(1-x^2)}\,dx$$

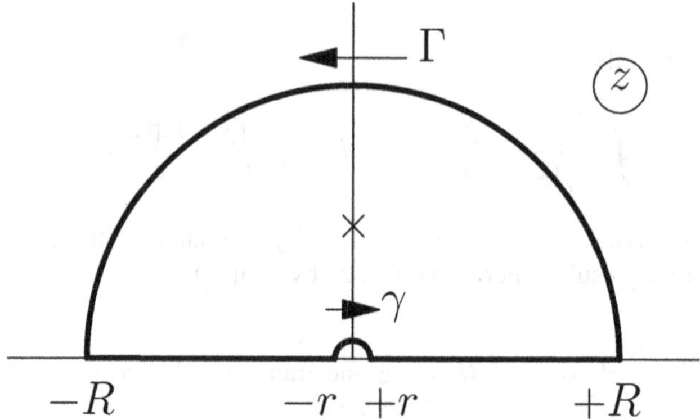

Fig. 2.1 See Problems 2.28–2.31

2.29
Evaluate the integrals

$$\int_{-\infty}^{+\infty} \frac{\sin x}{x(x-i)}\,dx\,; \quad \int_{-\infty}^{+\infty} \frac{\exp(ix)-\exp(-2ix)}{x(x-i)}\,dx$$

2.30
Evaluate the integral

$$\int_{-\infty}^{+\infty} \frac{\sin^2 x}{x^2}\,dx$$

Hint: Put $(\sin^2 x)/x^2 = \mathrm{Re}((1-\exp(2ix))/(2x^2))$ in order to have a simple pole on the real axis. This integral can also be evaluated in a different way: see Problem 3.9, q.(1).

2.31
Evaluate the following integrals, containing a Cauchy principal part due to the presence of singular points on the real axis (a doubt may arise in the calculation of the residue at $z = i$ in the third integral: one has indeed to perform the derivative of a function which contains a principal part in its definition. It can be seen, however, that one can safely evaluate the derivative simply replacing $\mathrm{P}(1/x)$ with $1/z$)

$$\mathrm{P}\int_{-\infty}^{+\infty} \frac{\exp(ix\pi)}{(x+1)(x+i)}\,dx\,; \quad \mathrm{P}\int_{-\infty}^{+\infty} \frac{\exp(-ix)}{x(x-i)^2(x+2i)}\,dx\,; \quad \mathrm{P}\int_{-\infty}^{+\infty} \frac{\exp(ix)}{x(x-i)^2}\,dx$$

2.32
Evaluate the integrals (use the closed contour in Fig. 2.2). *Hint*: $\cosh(x+\pi i) = -\cosh x$, and $\cos\big(a(x+\pi i)\big) = \cos ax\ \cosh a\pi + i \sin ax \sinh a\pi$, but the second

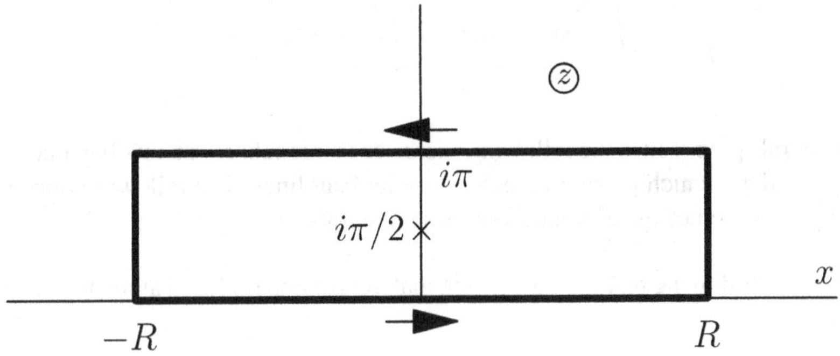

Fig. 2.2 See Problems 2.32, 2.39, 2.40

term does not contribute to the integral in the line at $z = i\pi$ because $\sin ax$ is an odd function.

$$\int_{-\infty}^{+\infty} \frac{x^2}{\cosh x}\, dx\,; \quad \int_{-\infty}^{+\infty} \frac{\cos(ax)}{\cosh x}\, dx \quad (a \in \mathbf{R})$$

2.33
Evaluate the integral (use the contour in Fig. 2.3)

$$\int_0^{+\infty} \exp(i\, x^2)\, dx$$

and deduce the integrals

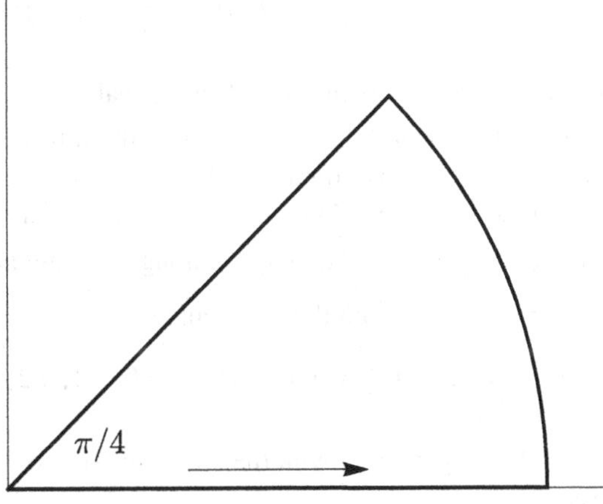

Fig. 2.3 See Problem 2.33

$$\int_0^{+\infty} \sin(x^2)\,dx = \int_0^{+\infty} \cos(x^2)\,dx$$

2.34

The integrals proposed in the following exercises of this subsection involve functions, as $\log z$ and z^α, which present branch points and cut lines. The following questions are in preparation of the evaluation of these integrals:

(1) Find the following residues, assuming that the cut line is placed along the positive real axis:

(a) residue at $z = \pm i$ of $f(z) = \dfrac{\sqrt{z}}{z \mp i}$ $\quad(b)$ residue at $z = \pm i$ of $f(z) = \dfrac{\log z}{z \mp i}$

(c) residue at $z = -1$ of $f(z) = \dfrac{\sqrt{z}}{z+1}$ $\quad(d)$ residue at $z = -i$ of $f(z) = \dfrac{z^{4/3}}{(z+i)^2}$

(e) residue at $z = -i$ of $f(z) = \dfrac{\sin(\pi\sqrt{2z})}{z+i}$

(f) residue at $z = -i$ of $f(z) = \dfrac{\exp(\pi\sqrt{2z})}{z+i}$

(2) Find the following residues

(a) residue at $z = i$ of $f(z) = \dfrac{\log(z^2-1)}{z-i}$ assuming that the cut is from $-\infty$ to -1 and from 1 to $+\infty$ in the real axis

(b) residue at $z = i$ of $f(z) = \dfrac{\sqrt{z^2-1}}{z-i}$ assuming that the cut is along the segment $[-1, 1]$ of the real axis

(c) residues at the (isolated) singular points of the function

$$f(z) = \frac{\sqrt{z-1}}{(z^2-1)(z^2+1)}$$

with the cut along the line $x \geq 1$ in the real axis

(3) Find the discontinuity presented by the following functions along the indicated branch cut (the discontinuity is defined as the value $f^+(x)$ of the function at the *upper* margin of the cut minus the value $f^-(x)$ at the *lower* margin):

(a) $f(z) = \sqrt{z^2-1}$ with the cut along the segment $[-1, 1]$ of the real axis

(b) $f(z) = \sqrt{\frac{z+1}{z-1}}$ with the same cut as in (a)

(c) $f(z) = (z-1)^\alpha$ with $\alpha \in \mathbf{R}$ $(\alpha \neq 0, \pm 1, \pm 2, \ldots)$ with the cut from 1 to $+\infty$ in the real axis

(d) $f(z) = \log(z^2-1)$ with the cut from $-\infty$ to -1 and from 1 to $+\infty$ in the real axis

(e) $f(z) = \log\dfrac{z+1}{z-1}$ with the same cut as in (c)

(*f*) $f(z) = \log \dfrac{z+1}{z-1}$ with the cut along the segment $[-1, 1]$ of the real axis

2.35

Evaluate the integrals (use the closed contour in Fig. 2.4, where the cut is indicated by a dashed line)

$$\int_0^\infty \frac{x^{\pm 1/2}}{1+x^2}\,dx\,; \quad \int_0^\infty \frac{x^{\pm 1/3}}{1+x^2}\,dx\,; \quad \int_0^\infty \frac{\sqrt{x}}{(x+i)^2}\,dx$$

2.36

Evaluate the integrals (use the closed contour as in the problem above)

$$\int_0^{+\infty} \frac{x^a}{(1+x)^2}\,dx \quad (-1 < a < 1)\,; \qquad \int_0^{+\infty} \frac{x^b}{1+x^3}\,dx \quad (-1 < b < 2)$$

(with $a \neq 0$; $b \neq 0$ and $\neq 1$, otherwise no cut line!)

2.37

(1) Integrating the functions

$$f(z) = \frac{\log z}{1+z+z^2}\,; \qquad f(z) = \frac{\log z}{1+z^3}$$

along a closed contour as in Fig. 2.4, obtain the integrals

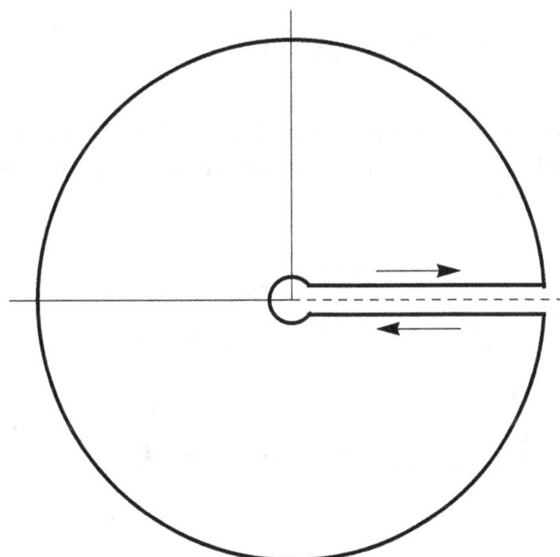

Fig. 2.4 See Problems 2.35, 2.36, 2.37

$$\int_0^{+\infty} \frac{1}{1+x+x^2}\,dx\,; \quad \int_0^{+\infty} \frac{1}{1+x^3}\,dx$$

(2) Integrating the function

$$f(z) = \frac{(\log z)^2}{1+z^2}$$

along the same closed contour, obtain the integrals (which can also be obtained by different methods, see next problem for the first integral, the second is elementary)

$$\int_0^{+\infty} \frac{\log x}{1+x^2}\,dx\,; \quad \int_0^{+\infty} \frac{1}{1+x^2}\,dx$$

2.38
Evaluate the integrals

$$\int_0^{+\infty} \frac{\log x}{1+x^2}\,dx\,; \quad \int_0^{+\infty} \frac{\log x}{1+x^4}\,dx$$

using for the first integral the closed contour shown in Fig. 2.5a. As suggested by the figure, the line along the positive real axis is chosen on the "upper margin" of the cut. *Hint*: if $x < 0$, then $\log x = \cdots$. For the second integral use the contour shown in Fig. 2.5b. *Hint*: if $y > 0$, then $\log(iy) = \log|y| + i\pi/2$

2.39
Evaluate the integrals

$$\int_{-\infty}^{+\infty} \frac{\exp(ax)}{1+\exp x}\,dx \quad (0 < a < 1)\,; \quad \int_0^{+\infty} \frac{1}{x^b(1+x)}\,dx \quad (0 < b < 1)$$

The first integral can be evaluated using a contour similar to that in Fig. 2.2, but with height $2\pi i$; for the second one, use the contour as in Fig. 2.4; see Problem 2.36.

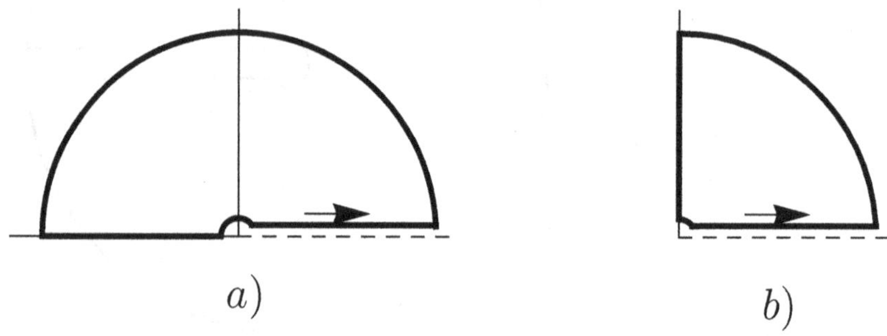

Fig. 2.5 See Problem 2.38

Putting $x' = \exp x$ the first integral is transformed into the second one, with $b = 1 - a$.

2.40
Evaluate the integral

$$\int_0^{+\infty} \frac{\log^2 x}{1+x^2}\, dx$$

Put $x' = \log x$ and use the same contour as in Fig. 2.2; see Problem 2.32.

2.41
Evaluate the integrals (use the closed contour as in Fig. 2.6). *Hint*: do not forget the residue at $z = \infty$; observe that the integration along the contour is performed in *clockwise* direction. For the first integral consider $\oint f(z)\, dz$ with $f(z) = \sqrt{z^2 - 1}$ and see Problem 2.34, q.(3) *(a)*

$$\int_{-1}^{+1} \sqrt{1-x^2}\, dx\,; \quad \int_{-1}^{1} \sqrt{\frac{1+x}{1-x}}\, dx\,; \quad \int_{-1}^{1} \frac{\sqrt{1-x^2}}{2-x}\, dx\,; \quad \int_0^{1} \frac{1+x^2}{\sqrt{x(1-x)}}\, dx$$

2.42
Evaluate the integral (use the closed contour in Fig. 2.7)

$$\int_{-\infty}^{+\infty} \frac{\log(a^2+x^2)}{1+x^2}\, dx\,, \qquad (a > 1)$$

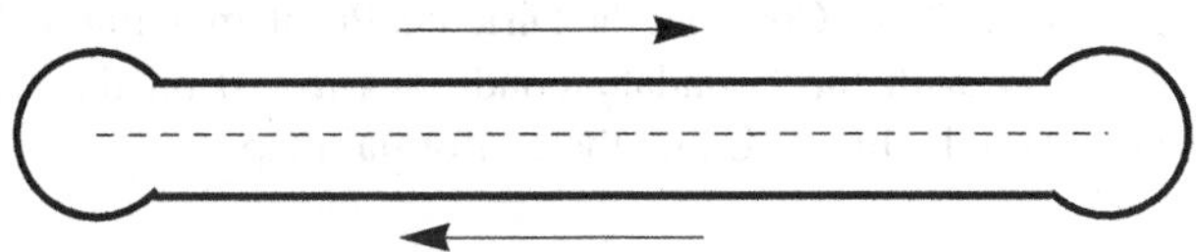

Fig. 2.6 See Problem 2.41

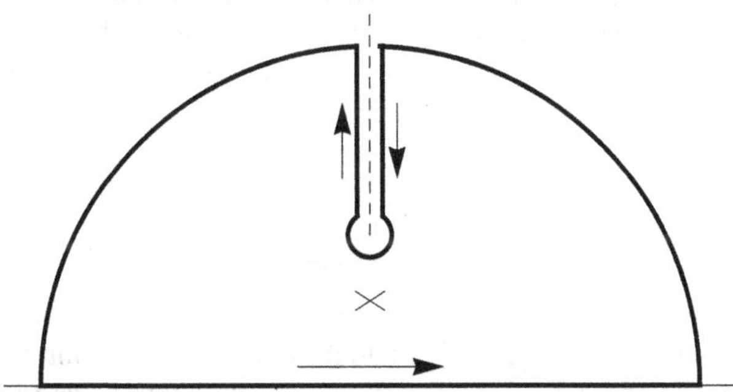

Fig. 2.7 See Problem 2.42

2.3 Harmonic Functions and Conformal Mappings

2.43

A harmonic function $u = u(x, y)$ can be interpreted, for instance—as well-known—as a 2-dimensional electric potential. Let $v = v(x, y)$ be its harmonic conjugate (unique apart from an additive constant), and let $f = u + iv$ be the corresponding analytic function. Then, the lines $u(x, y) =$ const. and $v(x, y) =$ const. represent the equipotential lines and respectively the lines of force of the electric field. Draw the lines $u(x, y) =$ const. and $v(x, y) =$ const. of the elementary analytic function $f(z) = z$ (trivial!), and of the functions

$$f(z) = z^2; \qquad f(z) = \log z \quad f(z) = \sqrt{z}$$

2.44

(1) Let D be the region in the complex plane z included between the positive real axis and the positive imaginary axis. Consider the Dirichlet Problem of finding a harmonic function $u(x, y)$ in D with the condition $u(x, y) = 0$ on the boundaries. Observing that the conformal map

$$z \to z' = \Phi(z) = z^2$$

maps D into ..., where the problem admits an elementary solution, solve the given Dirichlet Problem, and construct the lines $u(x, y) =$ const. and $v(x, y) =$ const.

(2) Consider now the region D in the complex plane z included between the positive real axis and the half straight line ℓ starting from the origin and forming an angle α with the real axis, see Fig. 2.8a. Consider the Dirichlet Problem of finding a harmonic function $u(x, y)$ in D with the boundary conditions $u = 0$ on the real axis and $u = u_0 =$ const $\neq 0$ on the line ℓ. Using the conformal map

$$z \to z' = \Phi(z) = \log z$$

which maps D into ..., where the problem admits an elementary solution, solve this Dirichlet Problem, and construct the lines $u(x, y) =$ const. and $v(x, y) =$ const.

2.45

(1) Show that the conformal map

$$z \to z' = \Phi(z) = \frac{i - z}{i + z}$$

transforms the half plane $y = \mathrm{Im}\, z \geq 0$ into the circle $|z'| \leq 1$. *Hint*: $|z'| = 1$ corresponds to the points z such that $|z - i| = |z + i|$, i.e., the points having equal distance from i and $-i$, which are the points of the real axis x, etc.

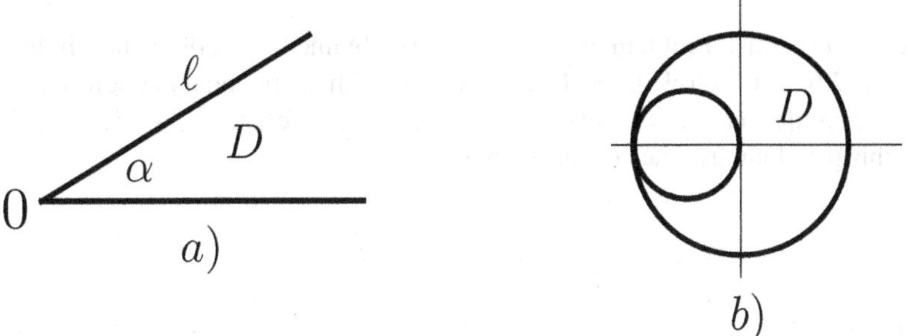

Fig. 2.8 See Problems 2.44 and 2.47

(2) Consider the Dirichlet Problem of finding the harmonic function $u(x, y)$ in the half plane $y \geq 0$ satisfying a given boundary condition on the x axis:

$$u(x, 0) = F(x)$$

Using the conformal map given in (1), with $z' = r' \exp(i\varphi')$, show that the problem becomes a Dirichlet Problem for the circle of radius $r' = 1$ centered at the origin of the plane z', where the boundary condition becomes

$$\widetilde{u}(1, \varphi') = F\big(\tan(\varphi'/2)\big) \equiv \widetilde{F}(\varphi'), \qquad -\pi < \varphi' < \pi$$

Hint: In the circumference $|z'| = r' = 1$ verify that $x = \tan(\varphi'/2)$.

(3) Using the result seen in (2) and recalling Sect. 1.1.3, solve the Dirichlet Problem for the half plane $y \geq 0$ if

$$F(x) = \frac{1}{1 + x^2}$$

Hint: $\widetilde{u}(r', \varphi') = (1 + r' \cos\varphi')/2 = \mathtt{Re}\,(1 + z')/2$, then ...

(4) The same as in (3) if

$$F(x) = \frac{x^2}{(1 + x^2)^2}$$

2.46

Show that the conformal map given in Problem 2.45, q.(1) transforms the line $z = x + i$ into the circumference $|z' + 1/2| = 1/2$, and the half plane $y = \mathtt{Im}\, z \geq 1$ into the interior of this circle. What is the image in the plane z' of the strip $0 \leq y \leq 1$ in the plane z under the transformation given above?

2.47

Solve the Dirichlet Problem in the region D obtained excluding the circle $|z + 1/2| \leq 1/2$ from the circle $|z| \leq 1$, see Fig. 2.8b, with the boundary conditions $u = 0$ on the circumference $|z| = 1$ and $u = 1$ on the circumference $|z + 1/2| = 1/2$. To solve this problem, use the conformal map

$$z \to z' = \Phi(z) = i\frac{1-z}{1+z}$$

which is just the *inverse* of the map considered in Problems 2.45 and 2.46, and which transforms D into ... (see Problem 2.46), where the problem becomes elementary. Verify explicitly that the solution satisfies the given boundary conditions.

2.48

(1) Show that the Dirichlet Problem for the half plane $\mathrm{Im}\, z = y \geq 0$ does *not* admit unique solution, but actually admits infinitely many solutions: verify for instance that the real part of the analytic functions iz, iz^2, etc., satisfy $\Delta_2 u = 0$ and the vanishing boundary condition $u(x, 0) = 0$. What is the most general solution of this problem?

(2) The existence of infinitely many solutions of the Dirichlet Problem in the half plane may appear surprising if compared with the Dirichlet Problem for the circle, recalling also that the two problems are connected by a conformal map (see Problem 2.45). To explain this fact, choose, e.g., the simplest solution $u(x, y) = y = \mathrm{Re}\,(-iz)$ to the Dirichlet Problem in the half plane $\mathrm{Im}\, z \geq 0$ satisfying the boundary condition $u(x, 0) = 0$, and find the corresponding solution $\widetilde{u}(r', \varphi')$ for the circle using the conformal map

$$z' \to z = \Psi(z') = i\frac{1-z'}{1+z'}$$

What is the singularity presented by this solution $\widetilde{u}(r', \varphi')$? Repeat calculations, for instance, for the solution in the half plane $u = xy = \mathrm{Re}\,(-iz^2/2)$, and verify that the same situation occurs. Conclude: how can one recover the uniqueness of the solution of the Dirichlet Problem? See also the discussion in Problem 3.112.

Chapter 3
Fourier and Laplace Transforms. Distributions

3.1 Fourier Transform in $L^1(\mathbf{R})$ and $L^2(\mathbf{R})$

There is some arbitrariness in the notation and definition of the Fourier transform. First of all, the independent variable of the functions to be transformed is often chosen to be the time $t \in \mathbf{R}$, and accordingly the independent variable of the transformed functions is the "frequency" $\omega \in \mathbf{R}$. This choice is due to the peculiar and characterizing physical interpretation of the Fourier transform as "frequency analysis". The notations frequently used in the following for the Fourier transform will then be:

$$\mathscr{F}\big(f(t)\big) = g(\omega) = \widehat{f}(\omega)$$

In the case that the function f to be transformed depends instead on the space position $x \in \mathbf{R}$, then its Fourier transform will depend on the "associate" physical variable $k = 2\pi/\lambda$, with usual notations, and we will write $\mathscr{F}\big(f(x)\big) = g(k) = \widehat{f}(k)$.

However, when no specific physical interpretation is involved and only the mathematical properties are concerned, the independent variable of the Fourier transform of a function $f(x)$ will be denoted either by k or by ω; sometimes a "neutral" variable $y \in \mathbf{R}$ will be used.

The definition of the Fourier transform (using the variables t and ω) for functions $f(t) \in L^1(\mathbf{R})$, which will be adopted is

$$\mathscr{F}\big(f(t)\big) = \int_{-\infty}^{+\infty} f(t)\exp(i\omega t)\,dt = g(\omega) = \widehat{f}(\omega)$$

and then for the inverse transformation

G. Cicogna, *Exercises and Problems in Mathematical Methods of Physics*, Undergraduate Lecture Notes in Physics,
https://doi.org/10.1007/978-3-030-59472-5_3

$$\mathscr{F}^{-1}\big(g(\omega)\big) = \frac{1}{2\pi}\int_{-\infty}^{+\infty} g(\omega)\exp(-i\omega t)\,d\omega = f(t)$$

Other definitions are possible: the factor $\exp(i\omega t)$ *in the first integral may be changed into* $\exp(-i\omega t)$ *and correspondingly the factor* $\exp(-i\omega t)$ *in the second one must be changed into* $\exp(i\omega t)$.

The factor $\frac{1}{2\pi}$ *in the second integral may be "distributed" into two factors* $\frac{1}{\sqrt{2\pi}}$ *in front to both integrals. As well-known, the Fourier transform and inverse Fourier transform defined in this way (using here the independent variables* x *and* y*)*

$$\widetilde{\mathscr{F}} = \frac{1}{\sqrt{2\pi}}\mathscr{F} \quad \text{i.e.} \quad \widetilde{\mathscr{F}}\big(f(x)\big) = \frac{1}{\sqrt{2\pi}}\int_{-\infty}^{+\infty} f(x)\exp(ixy)\,dx = g(y)$$

and

$$\widetilde{\mathscr{F}}^{-1} = \sqrt{2\pi}\,\mathscr{F}^{-1} \quad \text{i.e.} \quad \widetilde{\mathscr{F}}^{-1}\big(g(y)\big) = \frac{1}{\sqrt{2\pi}}\int_{-\infty}^{+\infty} g(y)\exp(-ixy)\,dy = f(x)$$

are unitary operators in the Hilbert space $L^2(\mathbf{R})$.

Notice that if $f(t) \in L^2(\mathbf{R})$ *but* $\notin L^1(\mathbf{R})$ *the correct notation for the Fourier transform would be*

$$\widehat{f}(\omega) = \mathtt{P}\int_{-\infty}^{+\infty} f(t)\exp(i\omega t)\,dt = \lim_{T\to+\infty}\int_{-T}^{T}\ldots$$

where the symbol $\mathtt{P}$ *denotes the Cauchy principal part of the integral (here, with respect to the infinity). For the sake of simplicity, the symbol* $\mathtt{P}$ *will be usually omitted. The same remark holds for the inverse Fourier transformation formula.*

Several examples of linear systems, where Fourier transform has important applications, will be proposed: these systems are characterized by an applied "input" which produces an answer, or "output", depending linearly on the input. The connection is often expressed by means of a linear differential equation, or—more in general—by specifying a Green function. In the second case, assuming that the independent variable is the time t, *denoting by* $a = a(t)$ *the input and by* $b = b(t)$ *the corresponding output, one can write*

$$b(t) = \int_{-\infty}^{+\infty} G(t,t')\,a(t')\,dt'$$

where $G = G(t,t')$ *is the Green function. In many cases one has* $G = G(t-t')$ *(thus the system has "time-invariant" properties), and* $b(t)$ *is expressed by a convolution product* $b(t) = \big(G * a\big)(t)$:

$$b(t) = \int_{-\infty}^{+\infty} G(t-t')\,a(t')\,dt' = \int_{-\infty}^{+\infty} G(\tau)\,a(t-\tau)\,d\tau$$

or equivalently $\widehat{b}(\omega) = \widehat{G}(\omega)\widehat{a}(\omega)$, *where* $\widehat{G}(\omega)$ *is sometimes called transfer function. Recall that the Green function is the answer to the Dirac delta input* $a(t) = \delta(t)$; *whereas* $b(t) = a(t)$ *for any* $a(t)$ *if* $G(t) = \delta(t)$ *(see Sect. 3.2). The Green function is said to be "causal" if* $G(t) = 0$ *for* $t < 0$. *This indeed guarantees that the answer* $b(t)$ *at any instant* t *depends only on the values of the input* $a(t')$ *at the "past" instants* $t' < t$. *Equivalently, assuming that there is some time* t_0 *such that the input* $a(t)$ *is equal to zero for any* $t < t_0$, *then also the corresponding solution, or answer,* $b(t)$ *is equal to zero for any* $t < t_0$. *In other words, the answer does not "precede" the input. For the case that the independent variable is not the time but a space variable* x, *see some comment before Problem 3.87.*

It is certainly useful to prepare a list of "basic" Fourier transforms and inverse transforms. Remark that many transforms can be obtained by means of elementary integration: e.g., if

$$f(t) = \theta(\pm t)\exp(\mp t), \quad \text{where } \theta(t) = \begin{cases} 0 & \text{for } t < 0 \\ 1 & \text{for } t > 0, \end{cases} \text{ then } \widehat{f}(\omega) = \frac{1}{1 \mp i\omega}$$

or if

$$\widehat{f}(\omega) = \begin{cases} 1 & \text{for } |\omega| < 1 \\ 0 & \text{for } |\omega| > 1 \end{cases}, \text{ then } f(t) = \frac{\sin t}{\pi t}$$

Other transforms require either the use of Jordan lemma, or—in some cases, more simply—the Fourier inversion theorem; e.g., using the above transforms, it is immediate to deduce

$$\mathscr{F}^{-1}\left(\frac{1}{\omega + i}\right) = \cdots \quad \text{and} \quad \mathscr{F}\left(\frac{\sin t}{t}\right) = \cdots$$

Elementary properties of Fourier transforms, as $\mathscr{F}\big(t^n f(t)\big) = \cdots$, *the translation theorems etc., can also be useful. To transform rational functions, one can either use Jordan lemma or the well-known decomposition of the function as a combination of simple fractions.*

3.1.1 Basic Properties and Applications

3.1
Find the Fourier transform $\widehat{f}(\omega)$ of the "quasi-monochromatic" wave with fixed frequency ω_0 in the time interval: $-t_0 < t < t_0$:

$$f(t) = \begin{cases} \exp(-i\omega_0 t) & \text{for} \quad |t| < t_0 \\ 0 & \text{for} \quad |t| > t_0 \end{cases}$$

Draw $|\widehat{f}(\omega)|$ assuming "large" t_0; show that the principal contribution to $\widehat{f}(\omega)$ is centered around ω_0, with maximum value proportional to t_0, according to the physical interpretation of the Fourier transform as "frequency analysis". See also Problem 3.15. For the limit case $t_0 \to \infty$, see Problem 3.25, q.(1).

3.2
Repeat the same calculations and considerations as in the above problem for these other types of "quasi-monochromatic" waves, with "large" a:

$$f_1(t) = \exp(-i\omega_0 t)\frac{a^2}{a^2 + t^2} \quad ; \quad f_2(t) = \exp(-i\omega_0 t)\,\exp(-t^2/a^2)$$

See also Problem 3.15. For the limit case $a \to \infty$, see Problem 3.25, q.(2).

3.3
Find the following Fourier and inverse Fourier transforms

$$\mathscr{F}\left(\frac{1}{t^2 + t + 1}\right) \quad ; \quad \mathscr{F}\left(\frac{t}{(t^2+1)^2}\right) \quad ; \quad \mathscr{F}\left(\frac{\cos t}{1 + t^2}\right) \quad ;$$

$$\mathscr{F}^{-1}\left(\frac{\omega \sin \omega}{1 + \omega^2}\right) \quad ; \quad \mathscr{F}^{-1}\left(\frac{1}{(\omega \pm i)^3}\right) \quad ; \quad \mathscr{F}^{-1}\left(\frac{\cos \omega}{(\omega \pm i)^4}\right)$$

3.4
(1) Consider the first-order linear non-homogeneous ordinary differential equation (ODE) for the unknown $x = x(t)$ (where $f(t) \in L^2(\mathbf{R})$ is given)

$$\dot{x} + x = f(t), \qquad x = x(t)$$

Introducing the Fourier transforms $\widehat{f}(\omega)$ and $\widehat{x}(\omega)$, obtain the solution $x(t)$ in the form of convolution product with a Green function $G(t)$, where $f(t)$ is the input and $x(t)$ the output: $x(t) = \big(G * f\big)(t)$. Find and draw the function $G(t)$. Explain why one obtains only one solution, and not ∞^1 solutions, as expected from general theory of first-order ODE's.

(2) The same questions for the equation

$$\dot{x} - x = f(t)\,, \qquad x = x(t)$$

The Green function is the answer to the Dirac delta input $\delta(t)$: so, one could expect that this answer should be zero for $t < 0$, i.e., that it should not *precede* the input: one of the Green functions obtained in (1–2) does not respect this property, why? (A first answer is essentially the same as that concerning the "absence" of solutions seen above, but more complete comments will be presented in the next sections in connection with the notion of "causality".)

(3) The equation of an electric series circuit of a resistance R and an inductance L

$$L\frac{dI}{dt} + RI = V(t)$$

where $I = I(t)$ is the current and $V = V(t)$ the applied voltage, has the same form as the equation in (1). The Fourier transform $\widehat{G}(\omega)$ has here an obvious physical interpretation ..., but there is apparently a "wrong" sign compared with the usual impedance formula $Z(\omega) = R + i\omega L$: why?

3.5

Using Fourier transform, find explicitly the solution $x(t)$ of the equations given in questions (1) and (2) of the above problem in the cases[1]

$$f(t) = \theta(t)\exp(-ct) \quad \text{and} \quad f(t) = \exp(-c|t|)\,, \qquad c > 0$$

with $c \neq 1$ and with $c = 1$.

3.6

(1) Consider the ODE

$$a\ddot{x} + b\dot{x} + cx = f(t)\,, \qquad x = x(t)$$

where the real constants a, b, c are > 0. Notice that this is, for instance, the equation of the motion of a particle of mass $m = a$, subjected to an elastic force $-c\,x$, to a viscous damping $-b\,\dot{x}$ and to an external time-dependent force $f(t)$. Find the Fourier transform $\widehat{G}(\omega)$ of the Green function $G(t)$ of this equation; show that all the singularities in the complex plane ω of $\widehat{G}(\omega)$ lie in the inferior half-plane $\omega'' = \mathrm{Im}\,\omega < 0$, with 3 possibilities: 2 single poles in $\omega_{1,2} = \pm\omega' - i\,\omega''$, 2 single poles in the imaginary axis, and a double pole. Find and draw $G(t)$ in each one of these cases. Verify also that in all cases the Green function is causal: $G(t) = 0$ when $t < 0$. (For the case $b = 0$, see Problem 3.78; for the case $c = 0$, see Problem 3.80; for the case $b = c = 0$, see Problem 3.79.)

[1] Direct calculation of the convolution product $(G * a)(t)$ is often not easy; it may be preferable to obtain first $\mathcal{F}(f(t)) = \widehat{f}(\omega)$ and then calculate $x(t)$ by inverse Fourier transform: $x(t) = \mathcal{F}^{-1}(\widehat{G}(\omega)\widehat{f}(\omega))$

(2) Find $G(t)$ for the ODE given in (1) choosing $a = 1/2,\ b = -1,\ c = 1$.

3.7

(1) Find and draw the Green function $G(t)$ by calculating the inverse Fourier transform in each one of the following four cases. What of these are causal, i.e. $G(t) = 0$ for $t < 0$?

$$\widehat{G}(\omega) = \frac{\exp(\pm i\omega)}{1 \pm i\omega}$$

(2) A Green function $G(t)$ has the following Fourier transform:

$$\widehat{G}(\omega) = \frac{\exp(\pm i\omega)}{P(\omega)}$$

where $P(\omega)$ is a polynomial with real coefficients (with no zero in the real axis). Can the Green function be causal?

(3) If the Fourier transform of a Green function is a *real* function $\widehat{G}(\omega)$, can the Green function be causal? And if the Fourier transform has the form $\widehat{G}(\omega) = \exp(i\omega)\widehat{G}_1(\omega)$ where $\widehat{G}_1(\omega)$ is real?

3.8

(1) Let $f(x) \in L^1(\mathbf{R}) \cap L^2(\mathbf{R})$. Using Fourier transform, give another proof (see Problem 1.4) that the subspace of the functions satisfying

$$\int_{-\infty}^{+\infty} f(x)\, dx = 0$$

is dense in $L^2(\mathbf{R})$. The same for the functions such that

$$\int_{-\infty}^{+\infty} f(x) \sin qx\, dx = 0 \qquad \text{for any } q \in \mathbf{R}$$

Hint: Observe that $\int_{-\infty}^{+\infty} f(x)\, dx = \widehat{f}(0)$, etc.

(2) Let $f(x)$ be such that $x^n f(x) \in L^1(\mathbf{R}) \cap L^2(\mathbf{R})$ for all integer $n \geq 0$: using Fourier transform show that the subspace of the these functions is dense in $L^2(\mathbf{R})$ (cf. Problem 1.3).

(3) Let $f(x)$ be such that $x^n f(x) \in L^1(\mathbf{R}) \cap L^2(\mathbf{R})$ for all integer $n \geq 0$ and

$$\int_{-\infty}^{+\infty} x^n\, f(x)\, dx = 0, \qquad \forall n \geq 0$$

using Fourier transform show that the set of the these functions is dense in $L^2(\mathbf{R})$.

3.9

To solve the following questions, recall that $\mathscr{F}\big((\sin x)/x\big) = \cdots$ and use Parseval identity for Fourier transform.

(1) Calculate (for a different procedure, based on integration in the complex plane, see Problem 2.30)

$$\int_{-\infty}^{+\infty} \frac{\sin^2 x}{x^2}\, dx$$

(2)(*a*) Consider the set of functions (recall also Problem 1.43, q.(2), observing that the Fourier transform is, apart from the coefficient $\sqrt{2\pi}$, an unitary operator)

$$f_n(x) = \frac{d^n}{dx^n}\frac{\sin x}{x}\,, \qquad n = 0, 1, 2, \ldots\,,\ x \in \mathbf{R}$$

show that $f_n(x)$ are not a complete set in $L^2(\mathbf{R})$. Characterize the functions $h(x) \in L^2(\mathbf{R})$ such that $(h, f_n) = 0,\ \forall n = 0, 1, 2, \ldots$

(*b*) Find $\| f_n(x)\|_{L^2(\mathbf{R})}$ and verify that $f_n(x) \to 0$, as $n \to \infty$, in the norm $L^2(\mathbf{R})$.

(3)(*a*) Show that the set of functions

$$g_n(x) = \frac{\sin x}{x - n\pi}\,, \qquad n \in \mathbf{Z},\ x \in \mathbf{R}$$

is an orthogonal but not a complete set in $L^2(\mathbf{R})$. Characterize the functions $h(x) \in L^2(\mathbf{R})$ such that $(h, g_n) = 0,\ \forall n \in \mathbf{Z}$.

(*b*) Show that the sequence $g_n(x)$ converges to 0 in weak $L^2(\mathbf{R})$ sense, not in norm.

3.10

(1) Consider the following integral $I(a)$ as a scalar product, as indicated

$$I(a) = \int_{-\infty}^{+\infty} \frac{\exp(i\omega a) - 1}{i\omega(1 + i\omega)}\, d\omega = \left(\frac{1}{1 - i\omega}\,, \frac{\exp(i\omega a) - 1}{i\omega}\right), \qquad a > 0$$

recalling the inverse Fourier transforms of the functions appearing in the scalar product and applying Parseval identity, evaluate $I(a)$. Evaluate then the limit $\lim_{a\to+\infty} I(a)$.

(2) Using Parseval identity as before, evaluate

$$\lim_{a\to+\infty} \int_{-\infty}^{+\infty} \frac{\exp(i\omega a) - 1}{i\omega} \exp(-\omega^2)\, d\omega$$

3.11

(1) Calculate the two inverse Fourier transforms

$$f_1(t) = \mathscr{F}^{-1}\left(\frac{\exp(i\omega)}{1 - i\omega}\right) \quad \text{and} \quad f_2(t) = \mathscr{F}^{-1}\left(\frac{\exp(-i\omega)}{1 + i\omega}\right)$$

(2) Show that the second inverse Fourier transform in the above question can be immediately obtained from the first one observing that $\widehat{f}(-\omega) = \mathscr{F}(\ldots)$.
(3) By integration in the complex plane, evaluate the integral

$$I = \int_{-\infty}^{\infty} \frac{\exp(2i\omega)}{(1 - i\omega)^2}\, d\omega$$

(4) Check the result obtained above observing that $I = (\widehat{f_2}, \widehat{f_1})$, using Parseval identity and the results in (1).

3.12

A particle of mass $m = 1$ is subjected to an external force $f(t)$ and to a viscous damping; denoting by $v = v(t)$ its velocity, the equation of motion is then

$$\dot{v} + \beta v = f(t)\,, \qquad t \in \mathbf{R},\ \beta > 0$$

(1) Let $f(t) = \theta(t)\exp(-t)$ and $\beta \neq 1$. Using Fourier transform, find $\widehat{v}(\omega) = \mathscr{F}\big(v(t)\big)$ and $v = v(t)$.
(2) One has from (1) that $\lim_{t\to+\infty} v(t) = 0$, then the final kinetic energy $v^2(+\infty)/2$ of the particle is zero. This implies that the work W_f done by the force, i.e.

$$W_f = \int_L f\, dx = \int_{-\infty}^{+\infty} f(t)v(t)\, dt$$

(where L is the space covered by the particle) is entirely absorbed by the work W_β done by friction, which is

$$W_\beta = \int_L \beta v\, dx = \beta \int_{-\infty}^{+\infty} v^2(t)\, dt$$

Evaluate W_f and W_β using the given $f(t)$ and the expression of $v(t)$ obtained in (1), and verify that indeed $W_f = W_\beta$.
(3) Repeat the above check using Fourier transform: precisely, thanks to the Parseval identity, write $W_f = \big(f, v\big)$ and $W_\beta = \beta\big(v, v\big)$ as integrals in the variable ω in terms of the expressions of $\widehat{f}(\omega)$ and $\widehat{v}(\omega)$ obtained above, evaluate these integrals by integration in the complex plane ω and verify again that $W_f = W_\beta$.
(4) The same questions in the case $\beta = 1$.

The following problem extends this result to the case of general forces $f(t) \in L^1(\mathbf{R}) \cap L^2(\mathbf{R})$. Problems 3.75 and 3.76 deal with the case of vanishing damping $\beta \to 0^+$.

3.13
This is the same problem as the previous one, now with a *general* force $f(t) \in L^1(\mathbf{R}) \cap L^2(\mathbf{R})$ (and $f(t)$ real, of course):

(1) Find the Fourier transform $\widehat{G}(\omega)$ of the Green function of the equation. Show that $\widehat{v}(\omega) = \mathcal{F}\big(v(t)\big) \in L^1(\mathbf{R})$, which implies that $\lim_{t\to+\infty} v(t) = \cdots$ and then that the final kinetic energy $v^2(+\infty)/2$ of the particle is zero.

(2) *(a)* Using Parseval identity, write the work done by the force

$$W_f = \int_L f\,dx = \int_{-\infty}^{+\infty} f(t)v(t)\,dt = \big(f, v\big) = (2\pi)^{-1}\big(\widehat{f}, \widehat{v}\big)$$

(where L is the space covered by the particle) as an integral in the variable ω in terms of $\widehat{v}(\omega)$ and $\widehat{G}(\omega)$.

(b) Proceeding as in *(a)*, write the work W_β done by friction

$$W_\beta = \int_L \beta v\,dx = \beta\int_{-\infty}^{+\infty} v^2(t)\,dt = \beta\big(v, v\big)$$

as an integral containing $\widehat{v}(\omega)$; show then that $W_f = W_\beta$, as expected (it is clearly understood that $v(-\infty) = 0$). *Hint*: recall that $f(t)$ is real, and the same is for $v(t)$, therefore $\widehat{v}^*(\omega) = \cdots$.

3.14
Consider a linear system described by a Green function $G(t)$, with input $a(t)$ and output $b(t)$ related by the usual rule $b(t) = \big(G * a\big)(t)$. Let $a(t) \in L^2(\mathbf{R})$.

(1) Assume $G(t) \in L^1(\mathbf{R})$. Show that $b(t) \in L^2(\mathbf{R})$ and find a constant C such that $\|b(t)\|_{L^2(\mathbf{R})} \le C\,\|a(t)\|_{L^2(\mathbf{R})}$.

(2) Assume $G(t) \in L^2(\mathbf{R})$. Show that $b(t)$ is a continuous function vanishing as $|t| \to \infty$.

(3) Assume that the Fourier transform $\widehat{G}(\omega) = \mathcal{F}\big(G(t)\big) \in L^2(\mathbf{R})$ and has a compact support. Show that $b(t)$ is continuous and differentiable (how many times?). Can one expect that also $b(t)$ has compact support?

3.15
The "classical uncertainty principle" states that, given any function $f(t)$ (under some obvious regularity assumptions, see (1) below), one has

$$\Delta t\,\Delta\omega \ge 1/2$$

where the quantities Δt and $\Delta\omega$ are defined by

$$\Delta t^2 = \frac{1}{\|f(t)\|^2}\int_{-\infty}^{+\infty}\big(t-\overline{t}\big)^2|f(t)|^2\,dt \quad , \quad \overline{t} = \frac{1}{\|f(t)\|^2}\int_{-\infty}^{+\infty} t\,|f(t)|^2\,dt$$

$$\Delta\omega^2 = \frac{1}{\|\widehat{f}(\omega)\|^2}\int_{-\infty}^{+\infty}(\omega-\overline{\omega})^2|\widehat{f}(\omega)|^2\,d\omega \ , \quad \overline{\omega} = \frac{1}{\|\widehat{f}(\omega)\|^2}\int_{-\infty}^{+\infty}\omega\,|\widehat{f}(\omega)|^2\,d\omega$$

(1) To show this result, assume for simplicity $\overline{t}=\overline{\omega}=0$ (this is not restrictive), assume $f(t)\in L^2(\mathbf{R})$, $tf(t)\in L^2(\mathbf{R})$, $\omega\widehat{f}(\omega)\in L^2(\mathbf{R})$ and finally that $\omega\,\widehat{f}(\omega)$ vanishes as $|\omega|\to\infty$; then verify and complete the steps of the following calculation:

$$0 \le \left\|\frac{\omega}{2\Delta\omega^2}\widehat{f}(\omega)+\frac{d\widehat{f}(\omega)}{d\omega}\right\|^2 = \cdots$$

$$= \frac{\|\widehat{f}(\omega)\|^2}{4\Delta\omega^2}+\|\mathscr{F}\big(itf(t)\big)\|^2+\frac{1}{2\Delta\omega^2}\int_{-\infty}^{+\infty}\omega\frac{d}{d\omega}\Big(\widehat{f}(\omega)\widehat{f}^*(\omega)\Big)\,d\omega = \cdots$$

$$= 2\pi\,\|f(t)\|^2\left(\Delta t^2-\frac{1}{4\Delta\omega^2}\right)$$

(2) Observing that the "minimum uncertainty", i.e. $\Delta t\,\Delta\omega = 1/2$, occurs when the quantity appearing in the first line above is zero, show that the minimum is verified when $\widehat{f}(\omega)=\cdots$ and then (now with generic $\overline{t}$ and $\overline{\omega}$, not necessarily zero) when

$$f(t) = c\,\exp\big(-\alpha^2(t-\overline{t})^2/2\big)\exp(-i\overline{\omega}t)$$

Find in this case Δt and $\Delta\omega$.

(3) Give an estimation of the spatial length $\Delta x = c\,\Delta t$ of the wave packet of a red light with $\lambda\simeq 7000$ Å and $\Delta\omega/\omega=\Delta\lambda/\lambda\simeq 10^{-6}$ (e.g., an atomic emission), and of a red laser wave with $\Delta\omega/\omega\simeq 10^{-12}$.

(4) Changing the variables t and ω into x and $k=2\pi/\lambda$ and using de Broglie principle $\lambda=h/p$, deduce the well-known Heisenberg uncertainty principle in quantum mechanics $\Delta x\,\Delta p\ge\hbar/2$.

3.1.2 Fourier Transform and Linear Operators in $L^2(\mathbf{R})$

This subsection is devoted to considering first examples of linear operators $T: L^2(\mathbf{R})\to L^2(\mathbf{R})$ which can be conveniently examined introducing their "Fourier transform" $\widehat{T}$ defined in this way: if

$$Tf(x)=g(x)\,, \qquad f(x),\ g(x)\in L^2(\mathbf{R})$$

then $\widehat{T}$ is the operator such that

$$\widehat{T}\,\widehat{f}(\omega) = \widehat{g}(\omega) \quad , \quad \text{i.e.} \quad \widehat{T} = \mathscr{F}\,T\,\mathscr{F}^{-1}.$$

Other more general examples of operators in the context of Fourier transforms and distributions will be considered in Sect. 3.2.2.

3.16
(1) Show that $\|\widehat{T}\| = \|T\|$.
(2) Assume that $\widehat{T}$ is a projection (on some subspace $H_1 \subset L^2(\mathbf{R})$); is the same true for T? on what subspace?

(3) Assume that $\widehat{T}$ admits an eigenvector $\varphi = \varphi(\omega)$ with eigenvalue λ; what information can be deduced for T?

(4) Assume that $\widehat{T}$ admits a orthonormal complete system of eigenvectors; does the same hold for T?

3.17
Find the operator $\widehat{T}$ in each one of the following cases

$$T\,f(x) = f(x-a) \quad \text{with} \quad a \in \mathbf{R} \quad ; \quad T\,f(x) = \frac{df}{dx} \quad ;$$

$$T\,f(x) = \int_{-\infty}^{+\infty} f(x-y)\,g(y)\,dy \equiv \big(f * g\big)(x) \quad \text{with} \quad g(x) \in L^1(\mathbf{R}) \cap L^2(\mathbf{R})$$

3.18
Consider the operator defined in $L^2(\mathbf{R})$

$$T\,f(t) = \int_{-\infty}^{+\infty} f(\tau)\frac{\sin(t-\tau)}{\pi(t-\tau)}\,d\tau$$

Introducing Fourier transform,

(1) Show that T is a projection: on what subspace?

(2) T has a clear physical interpretation in terms of the variable ω: explain!

(3) What differentiability properties can be deduced for the function $g(t) = T\,f(t)$? and about its behaviour as $|t| \to \infty$?

(4) Study the convergence as $n \to \infty$ of the sequence of operators

$$T_n\,f(t) = \int_{-\infty}^{+\infty} f(\tau)\frac{\sin\big(n(t-\tau)\big)}{\pi(t-\tau)}\,d\tau$$

3.19
Consider the operator defined in $L^2(\mathbf{R})$

$$T\,f(x) = \int_{-\infty}^{+\infty} f(y)\frac{1}{1+(x-y)^2}\,dy$$

Introducing Fourier transform,

(1) Find $\|T\|$

(2) Find $\mathtt{Ran}\,T$, specifying if $\mathtt{Ran}\,T = L^2(\mathbf{R})$ or at least is dense in it.

(3) For what $\rho \in \mathbf{R}$ does the operator $T - \rho I$ admit bounded inverse?

(4) What differentiability properties for the functions $g(x) = Tf(x)$ can be expected?

(5) If $\{f_n(x)\}$ is a complete set in $L^2(\mathbf{R})$, is the same true for the set $g_n = Tf_n$?

(6) Study the convergence as $a \to 0$ of the family of operators

$$T_a\,f(x) = \int_{-\infty}^{+\infty} f(y)\frac{a}{a^2+(x-y)^2}\,dy\,, \qquad a > 0$$

3.20

(1) Using Fourier transform, show that the family of operators T_a

$$T_a f(x) = f(x-a)\,, \qquad a \in \mathbf{R}$$

converges weakly to zero as $a \to \infty$.

(2) Show that T_a converges strongly to the identity operator I as $a \to 0$.

3.21

Introduce the slightly different definitions (in the factors $\sqrt{2\pi}$) of the Fourier transform $\widetilde{\mathscr{F}} = \frac{1}{\sqrt{2\pi}}\mathscr{F}$ and of the inverse Fourier transform $\widetilde{\mathscr{F}}^{-1} = \sqrt{2\pi}\mathscr{F}^{-1}$, i.e.,

$$\widetilde{\mathscr{F}}\big(f(x)\big) = \frac{1}{\sqrt{2\pi}}\int_{-\infty}^{+\infty} f(x)\exp(ixy)\,dx = g(y)$$

and

$$\widetilde{\mathscr{F}}^{-1}\big(g(y)\big) = \frac{1}{\sqrt{2\pi}}\int_{-\infty}^{+\infty} g(y)\exp(-ixy)\,dy = f(x)$$

It is known that $\widetilde{\mathscr{F}}$ (and $\widetilde{\mathscr{F}}^{-1}$ of course) are unitary operators in the Hilbert space $L^2(\mathbf{R})$: this follows from the general properties of the Fourier transform in $L^2(\mathbf{R})$: $\mathscr{F}$ is invertible and $\big(\widetilde{\mathscr{F}}(g), \widetilde{\mathscr{F}}(f)\big) = (f, g)$ from Parseval identity.

(1) Show that $\widetilde{\mathscr{F}}^2 = S$, where S is the parity operator: $S\,f(x) = f(-x)$, and then $\widetilde{\mathscr{F}}^4 = I$.

(2) Let T be the Hermite operator:

$$T = -\frac{d^2}{dx^2} + x^2$$

Show that $\widehat{T} = T$ and then $\widetilde{\mathscr{F}}T = T\widetilde{\mathscr{F}}$.

(3) Recall that the eigenfunctions of T are the Hermite functions $u_n = \exp(-\frac{x^2}{2})H_n(x)$ (where $H_n(x)$ are polynomials, $n = 0, 1, 2, \ldots$), and that the corresponding eigenvalues $\lambda_n = 2n + 1$ are non-degenerate: using then the result seen in (1) and (2), find the eigenfunctions and the eigenvalues of the operator $\widetilde{\mathscr{F}}$.

(4) Conclude: show that the Fourier operator $\widetilde{\mathscr{F}}$ is a linear combination of 4 projections; on what subspaces?

3.2 Tempered Distributions and Fourier Transforms

First of all, a remark concerning notations: the symbol T is used to denote distributions only in the present Introduction and in Sect. 3.2.1.

We will be almost exclusively concerned with the space of "tempered" distributions $\mathscr{S}'$, which is the "dual" space of the space $\mathscr{S}$ of the test functions $\varphi = \varphi(x)$ (i.e., the space of C^∞ functions rapidly vanishing as $|x| \to \infty$ with their derivatives); in other words, $\mathscr{S}'$ is the space of the continuous linear functionals $T : \mathscr{S} \to \mathbf{C}$. Instead of the "operatorial" notation $T(\varphi)$, we will adopt the notation $< T, \varphi >$, often used by physicists. The space $\mathscr{S}'$ can be considered as the "largest" space where Fourier transforms are introduced in a completely natural and well-defined way.

There are distributions which are associated to "ordinary" functions $u = u(x)$, defining $< T_u, \varphi >= \int_{-\infty}^{+\infty} u(x)\varphi(x)\,dx$, often writing simply $< u, \varphi >$ instead of $< T_u, \varphi >$. Notice that if $u(x) \in L^2(\mathbf{R})$ then $< u, \varphi >= (u^*, \varphi)$ is just a scalar product in $L^2(\mathbf{R})$, but $u(x)$ may be a much more general function, e.g., a polynomial or a trigonometric function: see the exercises below.

There is in $\mathscr{S}'$ the notion of convergence "in the sense of distributions": given a sequence $T_n \in \mathscr{S}'$ (or a family $T_a \in \mathscr{S}'$), one says that $T_n \to T$ in $\mathscr{S}'$ if for any test function $\varphi(x) \in \mathscr{S}$ one has $< T_n, \varphi > \to < T, \varphi >$. Accordingly, one can consider the convergence in $\mathscr{S}'$ of a sequence of functions $u_n(x)$, meaning the convergence of the distributions T_{u_n}. This is a new notion of convergence for sequences (or families) of functions, in addition (and to be compared) to the "old" notions, as, e.g., pointwise or uniform convergence, or—in the case $u_n(x) \in L^2(\mathbf{R})$—the convergence in the L^2 norm or in the weak L^2 sense. Recall that $u_n(x) \to u(x)$ in the sense of weak L^2 convergence if $\forall g \in L^2$, one has $(g, u_n) \to (g, u)$.

As said in the Introduction to Sect. 1.2, where sequences of linear operators were concerned, also in this section questions as "Study the convergence" or "Find the limit" of sequences (or families) of functions/distributions are "cumulative" questions, which require first, as obvious, to conjecture the possible limit, but also to specify in what sense the limit exists. By the way, also in this section, Lebesgue theorem will be a useful tool for examining convergence properties: see the introductory remarks to Sect. 1.2.

An important and useful property of the Fourier transform in $\mathscr{S}'$ is its continuity with respect to the notion of convergence in $\mathscr{S}'$: more precisely, one has that $T_n \to T$ in $\mathscr{S}'$ if and only if $\widehat{T}_n \to \widehat{T}$.

The notion of limit in $\mathscr{S}'$ is particularly relevant because it involves, for instance, the "approximation" of the Dirac delta $\delta(x)$ by means of "regular" (possibly C^∞) functions. The same holds for other distributions, as the derivatives of the delta, and, for instance, the important notion of Cauchy "Principal Part" $\mathtt{P}(1/x)$, defined by

$$< \mathtt{P}\frac{1}{x}, \varphi >= \lim_{\varepsilon \to 0+} \left(\int_{-\infty}^{-\varepsilon} + \int_{\varepsilon}^{+\infty} \frac{\varphi(x)}{x} dx \right), \qquad \varepsilon > 0$$

A function (or distribution) which will be used in the following is the "sign of x", i.e.

$$\mathtt{sgn}\, x = \begin{cases} -1 & \text{for } x < 0 \\ 1 & \text{for } x > 0 \end{cases} = \theta(x) - \theta(-x)$$

E.g., one has $\mathscr{F}(\mathtt{sgn}\, x) = 2i\, \mathtt{P}(1/\omega)$ and $\mathscr{F}(\mathtt{P}(1/x)) = \pi i\, \mathtt{sgn}(\omega)$, and $\mathscr{F}(\theta(x)) = \mathscr{F}\left(\frac{1}{2} + \mathtt{sgn}\, x\right) = i\, \mathtt{P}\left(\frac{1}{\omega}\right) + \pi\delta(\omega)$.

3.2.1 General Properties

3.22

Find the pointwise limit as $n \to \infty$ of the sequence of functions

$$f_n(x) = \begin{cases} n \sin nx & \text{for } 0 \le x \le \pi/n \\ 0 & \text{elsewhere} \end{cases}, \qquad n = 1, 2, \ldots; \ x \in \mathbf{R}$$

and then their limit in $\mathscr{S}'$. *Hint*: consider $f_n(x)$ as distributions, apply them to a generic test function $\varphi(x) \in \mathscr{S}$, i.e., $< f_n, \varphi >= \int_{-\infty}^{+\infty} \ldots$, perform a change of variable ... (cf. also Problem 1.1).

3.23

(1) *(a)* Find the limit in $\mathscr{S}'$ as $n \to \infty$ of the sequence of functions, with $x \in \mathbf{R}$,

$$f_n(x) = n\,\theta(x) \exp(-nx), \qquad n = 1, 2, \ldots$$

or, which is the same, the limit as $\varepsilon \to 0$ of $f_\varepsilon(x) = \varepsilon^{-1}\theta(x)\exp(-x/\varepsilon)$, $(\varepsilon > 0)$ (as in previous problem, consider the functions as distributions, apply them to a generic test function, etc.).

(b) Obtain again the limit using Fourier transform. *Hint*: the pointwise limit of the Fourier transforms can be easily obtained, and coincides, in this case, with the $\mathscr{S}'$-limit (why?), then ....

(2) The same questions *(a)* and *(b)* for the sequences of functions

$$f_n^{(1)}(x) = n\,\exp(-n|x|); \quad f_n^{(2)}(x) = n\,\exp(-x^2n^2); \quad f_n^{(3)}(x) = \frac{n}{1+n^2x^2}$$

(3) Find $\lim_{n\to\infty} g_n(x)$ in $\mathscr{S}'$ where $g_n(x) = n^3x\,\exp(-x^2n^2)$, both using one of limits obtained in (2) (notice that $g_n(x) \propto df_n^{(2)}/dx$), and using Fourier transform.

(4) Verify that, as well-known, the family of functions

$$u_\varepsilon(x) = \begin{cases} 1/(2\varepsilon) & \text{for } |x| < \varepsilon \\ 0 & \text{for } |x| > \varepsilon \end{cases}, \qquad x \in \mathbf{R}, \varepsilon > 0$$

tends to $\delta(x)$ as $\varepsilon \to 0$. Find then the Fourier transform $\widehat{u}_\varepsilon(y) = \mathscr{F}\big(u_\varepsilon(x)\big)$ and verify that indeed $\widehat{u}_\varepsilon(y) \to 1$.

(5)*(a)* The same as in (4) for this sequence of functions as $n \to \infty$ (which is a variant of (4) and is interesting because $\int_{-\infty}^{+\infty} u_n(x)\,dx = 0$)

$$u_n(x) = \begin{cases} n & \text{for } 0 < x < 1/n \\ -1 & \text{for } n < x < n+1 \\ 0 & \text{elsewhere} \end{cases}, \qquad x \in \mathbf{R},\ n = 1, 2, \dots$$

(b) What changes if the value of $u_n(x)$ in the interval $(n, n+1)$ is changed from -1 to $-n^\alpha$, or n^α (with $\alpha \in \mathbf{R}$)?

3.24

(1) To calculate the limit in $\mathscr{S}'$ as $n \to \infty$ of the sequence of functions

$$f_n(x) = \frac{\sin nx}{x}, \qquad x \in \mathbf{R}$$

recall that $\mathscr{F}\big((\sin nx)/x\big) = \cdots$, then find the limit (pointwise and $\mathscr{S}'$) of the Fourier transforms; therefore ....

(2) The limit in question (1) has been obtained as an application of the property "$f_n(x) \to \delta(x)$ if and only if the Fourier transforms $\widehat{f}_n(y) \to 1$". This property has been used in the previous problem and will be important in many of the following problems. Obtain now a direct proof that $\mathscr{F}\big(\delta(x)\big) = 1$ (and more in general

$\mathscr{F}(\delta(x-x_0)) = \exp(iyx_0))$ using the formal definition of Fourier transform of the distributions: $< \mathscr{F}T, \varphi > = < T, \widehat{\varphi} >$, where φ is any test function $\in \mathscr{S}$.

3.25

(1) Find the limit as $t_0 \to \infty$ of the "quasi-monochromatic" wave seen in Problem 3.1

$$f(t) = f_{t_0}(t) = \begin{cases} \exp(-i\omega_0 t) & \text{for} \quad |t| < t_0 \\ 0 & \text{for} \quad |t| > t_0 \end{cases}$$

and of its Fourier transform $\widehat{f_{t_0}}(\omega)$ (put $\omega - \omega_0 = x$ in $\widehat{f_{t_0}}(\omega)$, cf. previous problem). The physical interpretation is clear!

(2) The same questions for the limit as $a \to +\infty$ of the "quasi-monochromatic" waves seen in Problem 3.2.

3.26

(1) Let

$$u(t) = \begin{cases} 1 & \text{for} \quad 0 < t < 1 \\ 0 & \text{elsewhere} \end{cases}$$

The Fourier transform $\widehat{u}(\omega) = \mathscr{F}\big(u(t)\big)$ can be trivially obtained by elementary integration. Observing that one can write $u(t) = \theta(t) - \theta(t-1)$, one could also calculate $\widehat{u}(\omega)$ using the formula for $\mathscr{F}\big(\theta(t)\big)$: verify that the two results (seemingly different at first sight), after some simplifications, actually coincide, as expected!

(2) A similar exercise for finding the Fourier transform of

$$v(t) = \begin{cases} 0 \text{ for } 0 < t < 1 \\ 1 \text{ elsewhere} \end{cases} = 1 - u(t) = \theta(-t) + \theta(t-1)$$

3.27

Starting from the (well-known) Fourier transform of $f(x) = 1/(1+x^2)$, calculate the Fourier transform of $g(x) = x^2/(1+x^2)$ in these two different ways:

(a) writing

$$\frac{x^2}{1+x^2} = 1 - \frac{1}{1+x^2}$$

(b) using the rule $\mathscr{F}\big(x^2 f(x)\big) = -d^2\widehat{f}(k)/dk^2$.

3.28

Find the following Fourier and inverse Fourier transforms (D means derivation):

$$\mathscr{F}\big(|t|\big)\ ;\ \mathscr{F}^{-1}\big(|\omega|\big)\ ;\ \mathscr{F}\big(t\,\theta(t)\big)\ ;\ \mathscr{F}\big(\theta(t)\,\cos t\big)\ ;\ \mathscr{F}\left(\mathrm{P}\left(\frac{1}{t}\right)\frac{1}{1+t^2}\right)\ ;$$

$$\mathscr{F}^{-1}\left(D\,\mathrm{P}\left(\frac{1}{\omega}\right)\right)\ ;\ \mathscr{F}^{-1}\left(\mathrm{P}\frac{\omega}{\omega-1}\right)\ ;\ \mathscr{F}^{-1}\left(\frac{\exp(-i\omega)}{\omega+i}\mathrm{P}\left(\frac{1}{\omega}\right)\right)\ ;$$

$$\mathscr{F}\big(\exp(i|t|)\big)\ ;\ \mathscr{F}^{-1}\big(\omega^2\sin|\omega|\big)\ ;\ \mathscr{F}^{-1}\left(\mathrm{P}\left(\frac{1}{\omega}\right)\frac{1}{\omega\pm i}\right)\ ;$$

$$\mathscr{F}^{-1}\left(\mathrm{P}\frac{1}{\omega(\omega-1)}\right)=\mathscr{F}^{-1}\left(\mathrm{P}\left(\frac{1}{\omega-1}\right)-\mathrm{P}\left(\frac{1}{\omega}\right)\right)\ ;\ \mathscr{F}^{-1}\left(\mathrm{P}\left(\frac{\exp(i\omega)-1}{\omega^2}\right)\right)$$

3.29

(1) The Fourier transform $\widehat{f}(k)$ of the function

$$f(x)=\frac{1-\cos x}{x}\,,\qquad x\in\mathbf{R}$$

can be evaluated by integration in the complex plane using (with some care) Jordan lemma. Here, two alternative ways are proposed:

(a) put $g(x)=1-\cos x$ and evaluate first $\mathscr{F}\big(g(x)\big)=\widehat{g}(k)$; notice on the other hand that $\widehat{g}(k)=\mathscr{F}\big(xf(x)\big)=-i\,d\widehat{f}(k)/dk$, which gives $\widehat{f}(k)$ by direct integration (one constant must then be fixed ..., recall that $\widehat{f}(k)$ must belong to $L^2(\mathbf{R})$).

(b) write

$$\mathscr{F}\big(f(x)\big)=\mathscr{F}\left(\mathrm{P}\left(\tfrac{1}{x}\right)-\tfrac{1}{2}\exp(ix)\mathrm{P}\left(\tfrac{1}{x}\right)-\tfrac{1}{2}\exp(-ix)\mathrm{P}\left(\tfrac{1}{x}\right)\right)$$

and use $\mathscr{F}\big(\mathrm{P}(1/x)\big)=\cdots$.

(2) The same for the function

$$F(x)=\frac{1-\cos x}{x^2}\,,\qquad x\in\mathbf{R}$$

(a) with $g(x)=1-\cos x$ as before, now $\widehat{g}(k)=\mathscr{F}\big(x^2F(x)\big)=-d^2\widehat{F}(k)/dk^2$ (two constants must be fixed to determine $\widehat{F}(k)$ ..., recall that as before $\widehat{F}(k)\in L^2(\mathbf{R})$ and is continuous (why?)).

(b) write

$$\mathscr{F}\big(f(x)\big)=\mathscr{F}\left(-D\,\mathrm{P}\left(\tfrac{1}{x}\right)+\tfrac{1}{2}\exp(ix)D\,\mathrm{P}\left(\tfrac{1}{x}\right)+\tfrac{1}{2}\exp(-ix)D\,\mathrm{P}\left(\tfrac{1}{x}\right)\right)$$

and use $\mathscr{F}\big(D\,\mathrm{P}(1/x)\big)=\cdots$ (clearly, $D=d/dx$ is the derivative of distributions).

(3) Observing that $f(x)=x\,F(x)$, verify that $d\widehat{F}(k)/dk=\cdots$.

3.30

Find the Fourier transform of the following distribution

$$\mathrm{P}\left(\frac{\sin x}{x-a}\right),\qquad a\in\mathbf{R}$$

For what values of $a \in \mathbf{R}$ can the symbol P be omitted? for these values of a, what is the support of the Fourier transform?

3.31

(1) Using Fourier transform, find the "fundamental limits"

$$\lim_{\varepsilon \to 0^+} \frac{1}{x \pm i\varepsilon} = \mathrm{P}\left(\frac{1}{x}\right) \mp \pi\, i\delta(x)\,, \qquad \varepsilon > 0$$

and deduce

$$\lim_{\varepsilon \to 0^+} \frac{x}{x^2+\varepsilon^2}\ , \quad \lim_{\varepsilon \to 0^+} \frac{\varepsilon}{x^2+\varepsilon^2}$$

(2) Find then

$$\lim_{\varepsilon \to 0^+} \int_{-\infty}^{\infty} \frac{\exp(-x^4)}{x \pm i\varepsilon}\, dx$$

3.32

Let

$$u_a(\omega) = \frac{\exp(ia\omega) - 1}{i\omega}\,, \qquad a > 0$$

(1) Find the inverse Fourier transform $\mathscr{F}^{-1}\big(u_a(\omega)\big)$ and $\lim\limits_{a\to\infty} u_a(\omega)$.

(2) Using the above result, find

$$\lim_{a\to\infty} \int_{-\infty}^{+\infty} u_a(\omega) \cos\omega\, \exp(-\omega^2)\, d\omega$$

(3) Considering $\varphi(\omega) = \omega/(1+\omega^2)$ as a test function, use again the result obtained in (1) to find

$$\lim_{a\to\infty} \int_{-\infty}^{+\infty} u_a(\omega) \frac{\omega}{1+\omega^2}\, d\omega$$

(4) Check the above result: evaluate the integral (either by integration in the complex plane or using inverse Fourier transform)

$$I_a = \int_{-\infty}^{+\infty} \frac{\exp(ia\omega) - 1}{1+\omega^2}\, d\omega\,, \qquad a > 0$$

then find $\lim\limits_{a\to\infty} I_a$ and compare with the result in (3).

3.33

Let

$$u_a(x) = \begin{cases} -1 & \text{for } x < a \\ 1 & \text{for } x > a \end{cases}\ , \qquad x \in \mathbf{R},\ a \in \mathbf{R}$$

(1) Find $\lim_{a\to+\infty} u_a(x)$.

(2) Find the Fourier transform $\widehat{u}_a(y) = \mathscr{F}\big(u_a(x)\big)$ and $\lim_{a\to+\infty} \widehat{u}_a(y)$.

(3) Using the result in (2), find

$$\lim_{a\to+\infty} \mathrm{P}\int_{-\infty}^{+\infty} \frac{\exp(iay)}{y} \exp(-y^2)\, dy$$

(4) Considering $\varphi(y) = 1/(y-2i)$ as a test function, use again the result obtained in (2) to find

$$\lim_{a\to+\infty} \mathrm{P}\int_{-\infty}^{+\infty} \frac{\exp(iay)}{y(y-2i)}\, dy$$

(5) To check the above result evaluate now the integral by integration in the complex plane

$$I_a = \mathrm{P}\int_{-\infty}^{+\infty} \frac{\exp(iay)}{y(y-2i)}\, dy$$

then find $\lim_{a\to+\infty} I_a$ and compare with the result in (4).

3.34

(1) Observing that

$$u_\varepsilon(x) = \frac{1}{(x-i\varepsilon)^2} = -\frac{d}{dx}\frac{1}{x-i\varepsilon}, \qquad \varepsilon > 0$$

find the Fourier transform $\widehat{u}_\varepsilon(\omega) = \mathscr{F}\big(u_\varepsilon(x)\big)$ and $\lim_{\varepsilon\to0^+} \widehat{u}_\varepsilon(\omega)$.

(2) Find $\lim_{\varepsilon\to0^+} u_\varepsilon(x)$ either from (1) via inverse Fourier transform, or applying the result in Problem 3.31, q.(1).

(3) Using the result obtained in (2), evaluate the limits

$$\lim_{\varepsilon\to0^+}\int_{-\infty}^{+\infty} \frac{\exp(-x^2)}{(x-i\varepsilon)^2} \quad \text{and} \quad \lim_{\varepsilon\to0^+}\int_{-\infty}^{+\infty} \frac{x\exp(-x^2)}{(x-i\varepsilon)^2}$$

3.35

Let T_a be the distribution

$$T_a = \frac{1}{a}\left(\mathrm{P}\frac{1}{x-a} - \mathrm{P}\frac{1}{x}\right), \qquad a > 0$$

(1) Find the Fourier transform $\widehat{T}_a$ of T_a.

(2) Find the limit $\widehat{T} = \lim_{a\to0} \widehat{T}_a$ and find $T = \mathscr{F}^{-1}\widehat{T}$.

(3) Evaluate $< T, \exp(-x^2) >$ and $< T, \sin x \exp(-x^4) >$.

3.36

Verify carefully the following identities; next, obtain from them other (perhaps simpler) identities using Fourier transform:

(*a*) $x\,D\,\mathrm{P}\frac{1}{x} = -\mathrm{P}\frac{1}{x}$ (*b*) $x^2 D\,\mathrm{P}\frac{1}{x} = -1$ (*c*) $x^2 D^2 \mathrm{P}\frac{1}{x} = 2\,\mathrm{P}\frac{1}{x}$

3.37

Let $C(f)$ be the convolution product

$$C(f) = \int_{-\infty}^{+\infty} \exp(-|x-y|)\,\mathrm{sgn}(x-y)\,f(y)\,dy$$

Using Fourier transform, find the most general solution $f(x)$ of the following equations

(*a*) $C(f) = \delta(x)$ (*b*) $C(f) = x\exp(-x^2)$

(*c*) $C(f) = (i/2)\,f$ (*d*) $C(f) = 2i\,f$

3.38

Let $h_a(x) = \sin ax/(\pi x),\ a > 0,\ x \in \mathbf{R}$, and consider the convolution product

$$g_a(x) = \big(h_a * f\big)(x) = \int_{-\infty}^{+\infty} f(x-y)h_a(y)\,dy$$

Use Fourier transform.

(1) Let $f(x) \in L^2(\mathbf{R})$: show that $g_a(x)$ is infinitely differentiable and $\in L^2(\mathbf{R})$; find $\lim_{a\to\infty} g_a(x)$.

(2) Let $f(x) = \mathrm{P}(1/x)$: find $\mathscr{F}\big(g_a(x)\big)$ and $\lim_{a\to\infty} g_a(x)$.

(3) Let $f(x) = \delta'(x)$: find $\mathscr{F}\big(g_a(x)\big)$ and $\lim_{a\to\infty} g_a(x)$.

3.39

(1) By integration in the complex plane evaluate

$$I_a = \mathrm{P}\int_{-\infty}^{+\infty} \frac{\exp(iat)}{t(t-x)}\,dt\,, \qquad x \in \mathbf{R},\ a > 0$$

(2) Find the Fourier transform $\widehat{F}_a(\omega)$ of the convolution product

$$F_a(x) = \mathrm{P}\frac{1}{x} * \frac{\sin ax}{\pi x}\,, \qquad a > 0$$

What properties of $F_a(x)$ can be deduced from its transform: is $F_a(x)$ a bounded function? continuous and differentiable (how many times?), is $F_a(x) \in L^1(\mathbf{R}) \cap L^2(\mathbf{R})$?

(3) *(a)* check the answers given in (2) observing that $F_a(x) = -(1/\pi)\mathtt{Im}(I_a)$

(b) it is also easy to obtain $F_a(x)$ directly, evaluating the inverse Fourier transform of the function $\widehat{F}_a(\omega)$ obtained in (2).

(4) Find $\lim_{a\to+\infty} F_a(x)$ and conclude observing that the above results provide another approximation of the distribution $\mathtt{P}(1/x)$ with a C^∞ function (for a simpler approximation see Problem 3.31, q.(1)).

3.40

(1) Find the inverse Fourier transform $f_\varepsilon(t)$ of

$$\widehat{f_\varepsilon}(\omega) = \frac{1}{\omega + i\varepsilon}\frac{1}{\omega - i}, \qquad \varepsilon > 0$$

and then evaluate the limit (in $\mathscr{S}'$) $\lim_{\varepsilon\to 0^+} f_\varepsilon(t)$

(2) Exchange the operations: first evaluate $\lim_{\varepsilon\to 0^+} \widehat{f_\varepsilon}(\omega)$ and then find the inverse Fourier transform. The results should coincide! (why?)

3.41

(1) *(a)* Find the inverse Fourier transform $f_\varepsilon^{(+)}(t)$ of

$$g_\varepsilon^{(+)}(\omega) = \frac{1}{1 - (\omega + i\varepsilon)^2}, \qquad \varepsilon > 0$$

and then the limit (in $\mathscr{S}'$, of course) $F^{(+)}(t) = \lim_{\varepsilon\to 0^+} f_\varepsilon^{(+)}(t)$

(b) the same for the inverse Fourier transform $f_\varepsilon^{(-)}(t)$ of

$$g_\varepsilon^{(-)}(\omega) = \frac{1}{1 - (\omega - i\varepsilon)^2}, \qquad \varepsilon > 0$$

and for $F^{(-)}(t) = \lim_{\varepsilon\to 0^+} f_\varepsilon^{(-)}(t)$.

(2) Check the results in (1) evaluating first the limits $G^{(\pm)}(\omega) = \lim_{\varepsilon\to 0^+} g_\varepsilon^{(\pm)}(\omega)$ and then the inverse Fourier transforms $F^{(\pm)}(t)$ of $G^{(\pm)}(\omega)$. *Hint:* find first the following four limits (see Problem 3.31, q.(1))

$$\lim_{\varepsilon\to 0^+} \frac{1}{1 \pm \omega \pm i\varepsilon}$$

(3) Do the functions $F^{(+)}(t)$ and $F^{(-)}(t)$ coincide? See also next problem.

3.42

(1) The same questions as in the above problem for the functions (apparently similar to those in the problem above)

$$g_\varepsilon^{(+)}(\omega) = \frac{1}{(1+i\varepsilon)^2-\omega^2} \quad \text{and} \quad g_\varepsilon^{(-)}(\omega) = \frac{1}{(1-i\varepsilon)^2-\omega^2}, \qquad \varepsilon > 0$$

(2) Do the functions $F^{\pm}(t)$ coincide with those obtained in the problem above?
(3) Another possibility: Find the inverse Fourier transform $H(t)$ of

$$\widehat{H}(\omega) = \mathrm{P}\left(\frac{1}{1-\omega^2}\right) = \mathrm{P}\left(\frac{-1/2}{\omega-1}\right) + \mathrm{P}\left(\frac{1/2}{\omega+1}\right)$$

Does this $H(t)$ coincide with some of the functions obtained above?

3.43
(1) Consider the sequence of functions

$$f_n(x) = n\,\theta(x)\exp(-xn) \quad \text{and} \quad F_n(x) = \theta(x)\big(1-\exp(-xn)\big), \quad n = 1, 2, \ldots$$

Find $\widehat{f_n}(\omega) = \mathscr{F}\big(f_n(x)\big)$ and $\widehat{F}_n(\omega) = \mathscr{F}\big(F_n(x)\big)$.
(2) Show that $F_n'(x) = f_n(x)$ and, using the formula $\mathscr{F}\big(F'(x)\big) = \ldots$, obtain again $\widehat{f_n}(\omega)$ from $\widehat{F}_n(\omega)$.
(3) Find $\lim_{n\to\infty} f_n(x)$ and $\lim_{n\to\infty} F_n(x)$; verify that $\lim_{n\to\infty} f_n(x) = \dfrac{d}{dx}\Big(\lim_{n\to\infty} F_n(x)\Big)$.

3.44
Consider the family of functions

$$\widehat{f_a}(\omega) = i\,\frac{\exp(ia\omega)+\exp(-ia\omega)-2}{a^2\omega}, \qquad a > 0$$

(1) Find the pointwise limit $\lim_{a\to 0}\widehat{f_a}(\omega)$. Does the limit exist in $L^2(\mathbf{R})$, in $\mathscr{S}'$?
(2) Recalling that $\mathscr{F}^{-1}\big(\mathrm{P}(1/\omega)\big) = \cdots$, find and draw the inverse Fourier transform

$$f_a(x) = \mathscr{F}^{-1}\big(\widehat{f_a}(\omega)\big)$$

Why can the symbol P be omitted in the above $\widehat{f_a}(\omega)$?
(3) Using (1), find $\lim_{a\to 0} f_a(x)$.
(4) Considering $f_a(x)$ as distributions, choose $\varphi(x) = x\exp(-x^2)$ as test function: calculate $< f_a, \varphi >$ and $\lim_{a\to 0} < f_a, \varphi >$. Then check the answer given in (3).

3.45
Let $F_L(x)$ be the function

$$F_L(x) = \begin{cases} -L & \text{for } x \le -L \\ x & \text{for } |x| \le L \\ L & \text{for } x \ge L \end{cases}, \qquad L > 0$$

(1) Find $\widehat{F}_L(\omega) = \mathscr{F}\big(F_L(x)\big)$

(2) Find $f_L(x) = F'_L(x)$ and $\widehat{f}_L(\omega) = \mathscr{F}\big(f_L(x)\big)$; check the results obtained here and in (1) using the property $\widehat{f}_L(\omega) = \mathscr{F}\big(F'_L(x)\big) = -i\omega\mathscr{F}\big(F_L(x)\big)$.

(3) Find $\lim_{L\to\infty} \widehat{F}_L(\omega)$.

3.46

(1) Let $g(\omega)$ be the (given) Fourier transform of a function $f(x)$ and let $F(x)$ be such that $F'(x) = f(x)$. Show that, expectedly, the Fourier transform $G(\omega) = \mathscr{F}\big(F(x)\big)$ of $F(x)$ is determined by $g(\omega)$ apart from an additional term.

(2) Let

$$F(x) = \begin{cases} x & \text{for} \quad 0 < x < 1 \\ 0 & \text{elsewhere} \end{cases}$$

Find $f(x) = F'(x)$ (notice that $F(x)$ is not continuous, then $F'(x) = \cdots$) and find $g(\omega) = \mathscr{F}\big(f(x)\big)$; then deduce $G(\omega) = \mathscr{F}\big(F(x)\big)$ observing that the additional term can now be fixed thanks to the property that $G(\omega)$ must be a C^∞ function (why?).

(3) Confirm the above result observing that $F(x)$ can be written as $F(x) = x\,u(x)$ where

$$u(x) = \begin{cases} 1 & \text{for} \quad 0 < x < 1 \\ 0 & \text{elsewhere} \end{cases}$$

using $\mathscr{F}\big(x\,u(x)\big) = -i\,d\widehat{u}(\omega)/d\omega$.

(4) Let now

$$F_1(x) = \begin{cases} 0 & \text{for} \quad x < 0 \\ x & \text{for} \quad 0 < x < 1 \\ 1 & \text{for} \quad x > 1 \end{cases}$$

Observing that $F'_1(x) = u(x)$ (*not* $F'(x) = u(x)$, cf. question (1)), deduce $\mathscr{F}\big(F_1(x)\big)$; use now $F_1(x) = F(x) + \theta(x-1)$.

(5) Using the method proposed in (1), reconsider the previous problem and obtain $\widehat{F}_L(\omega) = \mathscr{F}\big(F_L(x)\big)$ starting from $\widehat{f}_L(\omega) = \mathscr{F}\big(F'_L(x)\big)$ (which is simpler to evaluate): the additional term can be fixed observing that the function $F_L(x)$ is an *odd* function (and the same is for $\widehat{F}_L(\omega)$) and $\delta(\omega)$ is *even*, therefore ....

3.47

(1) Observing that $\dfrac{d}{dx}\arctan x = \dfrac{1}{1+x^2}$ and that $\mathscr{F}\big(1/(1+x^2)\big) = \cdots$, use question (1) of the above problem to deduce $\mathscr{F}(\arctan x)$; the additional term can be fixed with the same argument as in q.(5) of the problem above. For a different way to obtain this Fourier transform, see next problem.

(2) Find $\lim_{n\to\infty} \arctan nx$ and $\lim_{n\to\infty} \mathscr{F}(\arctan nx)$.

3.48

Let the output $b(t)$ produced by a linear system when $a(t)$ is the applied input be given by

$$b(t) = \int_{-\infty}^{t} a(t')\,dt'$$

(1) Verify that the Green function of this system is $G(t) = \theta(t)$.

(2) Using the result $\mathscr{F}(\theta * f) = \cdots$, and choosing $a(t) = \exp(-t^2)$, find $\mathscr{F}\big(\text{erf}(t)\big)$, where $\text{erf}(t) = \int_{-\infty}^{t} \exp(-x^2)dx$.

(3) The same as in (2) choosing $a(t) = 1/(1+t^2)$: find again $\mathscr{F}(\arctan t)$ (cf. previous problem).

3.49

Let

$$f_0(t) = \begin{cases} 1 & \text{for } \ 0 < t < 1 \\ 0 & \text{elsewhere} \end{cases} \quad , \qquad t \in \mathbf{R}$$

and let $\widehat{f_0}(\omega)$ be its Fourier transform. Consider the series

$$\widehat{F_\varepsilon}(\omega) = \widehat{f_0}(\omega)\Big(1 + \exp(-\varepsilon)\,\exp(i\omega) + \exp(-2\varepsilon)\,\exp(2i\omega) + \cdots\Big)$$

$$= \widehat{f_0}(\omega) \sum_{n=0}^{\infty} \exp\big(-n(\varepsilon - i\omega)\big)\,, \qquad \varepsilon > 0$$

(1) Find and draw the inverse Fourier transform $F_\varepsilon(t) = \mathscr{F}^{-1}\big(\widehat{F_\varepsilon}(\omega)\big)$. *Hint*: use a translation theorem.

(2) Find the sum $\widehat{F_\varepsilon}(\omega)$ of the series.

(3) Find $\|\widehat{F_\varepsilon}(\omega)\|^2_{L^2} = 2\pi \, \|F_\varepsilon(t)\|^2_{L^2}$

(4) Find $\lim_{\varepsilon \to 0} \widehat{F_\varepsilon}(\omega)$.

3.50

(1) It is well-known that in general the product of distributions cannot be defined. E.g., one could try to define $\delta^2(x)$ starting with the product of some sequences of "regular" functions $u_n(x)$ which approximate $\delta(x)$. Several examples of such functions are proposed in the first problems of this subsection. Show that using all these examples, the sequences $u_n^2(x)$ (or $u_\varepsilon^2(x)$) have no limit.

(2) Verify that no result is obtained also considering sequences as $u_n(x)\,\delta(x)$ (the limit depends on the approximating sequence $u_n(x)$ which has been chosen).

3.51

As the product of distributions, the convolution product is in general not defined. Consider

$$\sin x * \frac{\sin ax}{x} = \int_{-\infty}^{+\infty} \sin(x-y)\frac{\sin ay}{y}\,dy\,, \qquad a>0$$

and use Fourier transform. Is this convolution product defined for all a?

3.52

(1) Let

$$u_\varepsilon(x) = \exp(-\varepsilon|x|)\,, \qquad \varepsilon>0$$

Calculate the convolution product

$$v_\varepsilon(x) = u_\varepsilon(x) * \mathtt{sgn}\,x$$

(find first the Fourier transform $\widehat{v}_\varepsilon(\omega) = \cdots$ and then its inverse Fourier transform).
(2) Find the limits in $\mathscr{S}'$ as $\varepsilon \to 0^+$ of $u_\varepsilon(x)$ and of $v_\varepsilon(x)$.
(3) One could conjecture that a definition of the convolution $1 * \mathtt{sgn}\,x$ (or equivalently, apart some factor, of the product $\delta(\omega)\mathrm{P}(1/\omega)$) could be given approximating in $\mathscr{S}'$ the constant function 1 with $u_\varepsilon(x)$ and then defining $1 * \mathtt{sgn}\,x$ as the limit of $v_\varepsilon(x)$. Explain why this is not correct: repeat calculations in (1) and (2) now approximating 1 with

$$u_{\varepsilon_1,\varepsilon_2}(x) = \begin{cases} \exp(\varepsilon_1 x) & \text{for } x<0 \\ \exp(-\varepsilon_2 x) & \text{for } x>0 \end{cases}\,, \qquad \varepsilon_1,\ \varepsilon_2>0,\ \varepsilon_1 \neq \varepsilon_2$$

(the definition cannot depend on the approximation chosen …!)

3.53

(1) Verify the (well-known) identity $x\,\delta'(x) = -\delta(x)$. Next, show that

$$x^2\delta(x) = x^2\delta'(x) = 0\ ,\quad x^2\delta''(x) = 2\delta(x)$$

and verify that the Fourier transforms of these identities produce obvious results.
(2) Generalize: find

$$x^2\delta'''(x) = \cdots\ ,\quad x^m\delta^{(n)}(x) = \cdots$$

(3) *(a)* Verify that

$$h(x)\,\delta(x-x_0) = h(x_0)\,\delta(x-x_0)$$

where $h(x)$ is any function continuous in a neighborhood of x_0.
(b) Extend to

$$h(x)\,\delta'(x-x_0) = \cdots$$

(c) The Fourier transform $\mathscr{F}(x-1)$ can be evaluated in two ways:

$$\mathcal{F}(x-1) = \mathcal{F}(x) - \mathcal{F}(1) \quad \text{and} \quad \mathcal{F}(x-1) = \exp(i\omega)\,\mathcal{F}(x)$$

as an application of *(b)*, verify that the two results coincide.

In the Problems 3.54–3.58 the independent variable is denoted by y, to avoid confusion with the notations used in other problems where the results obtained here are applied.

3.54

(1) Without using Fourier transform, find the most general distributions $T \in \mathcal{S}'$ which satisfy each one of the following equations:

$$(a)\ y\,T = 0\,; \quad (b)\ y\,T = 1\,; \quad (c)\ y^2 T = 0\,; \quad (d)\ y^2 T = 1\,; \quad (e)\ y^3 T = 0\,;$$

$$(f)\ y^3 T = 1\,; \quad (g)\ y\,T = \sin y; \quad (h)\ y^2\,T = 1 - \cos y; \quad (i)\ y\,T = \cos y$$

(2) Specify if there are solutions, among those found in (1), which belongs to $L^2(\mathbf{R})$ (clearly and more precisely: specify if there are distributions $T = T_u$ which are associated to functions $u(y) \in L^2(\mathbf{R})$).

3.55

The same questions as in the above problem for the equations:

$$(a)\ (y-1)\,T = 0\,; \quad (b)\ (y-1)\,T = 1\,; \quad (c)\ (y \pm i)T = 0\,; \quad (d)\ (y \pm i)T = 1;$$

$$(e)\ (y-\alpha)\,T = 0\,; \quad (f)\ (y-\alpha)\,T = 1 \quad \text{with } \alpha = a + ib,\ a, b \text{ real } \neq 0$$

3.56

The same questions as in Problem 3.54 for the equations:

$$(a)\ y(y-1)\,T = 0\,;\ (b)\ y(y-1)\,T = 1\,;\ (c)\ (y^2-1)\,T = 0\,;\ (d)\ (y^2-1)\,T = 1;$$

$$(e)\ y(y \pm i)\,T = 0\,;\ (f)\ y(y \pm i)\,T = 1\,;\ (g)\ (y^2+1)\,T = 0\,;\ (h)\ (y^2+1)\,T = 1$$

3.57

The same questions as in Problem 3.54 for the equations:

$$(a)\ (\sin y)\,T = 0\,; \quad (b)\ \big(1 + \exp(iy)\big)\,T = 0\,;$$

$$(c)\ (1-\cos y)\,T = 0\,; \quad (d)\ (y - \sin y)\,T = 0\,; \quad (e)\ \exp(-1/y^2)\,T = 0$$

3.58
Using Fourier transform find the distributions T which solve the following equations:

$$(a)\ y\,T = \delta(y)\,; \quad (b)\ (y-1)\,T = \delta(y)\,; \quad (c)\ y^2\,T = \delta(y)\,; \quad (d)\ y\,T = \mathrm{P}(1/y)$$

3.59
Using Fourier transform solve the following equations for the unknown function $u(x)$:

$$(a)\ u'(x) + u(x-\pi/2) = 0\,; \quad (b)\ u''(x) + u(x) = u(x-\pi) + u(x+\pi)\,;$$

$$(c)\ u(x-a) + 2u(x+2a) = 3u(x)\,, \qquad a > 0\,;$$

$$(d)\ 2u'(x) + u(x-1) - u(x+1) = 0$$

3.60
Using Fourier transform find the distributions T which solve the equation:

$$x\,DT + T = 0$$

(clearly DT is the derivative of T). Compare with the analogous ODE $x\,f' + f = 0$ for "elementary" functions $f = f(x)$.

3.61
Let $f_\varepsilon(x)$ be the function in $L^2(-\pi,\pi)$ defined by

$$f_\varepsilon(x) = \begin{cases} 1/(2\varepsilon) & \text{for } |x| < \varepsilon \\ 0 & \text{for } |x| > \varepsilon \end{cases}\,, \qquad 0 < \varepsilon < \pi$$

and evaluate its Fourier expansion in terms of the complete set $\{\exp(inx)\,,\, n \in \mathbf{Z}\}$. Consider then the periodic prolongation $\widetilde{f_\varepsilon}(x)$ to all $x \in \mathbf{R}$ of $f_\varepsilon(x)$ with period 2π. Evaluate the limit as $\varepsilon \to 0^+$ of $f_\varepsilon(x)$ and of its Fourier expansion (notice that this expansion is automatically periodic; explain why these limits can be safely performed) to obtain the Fourier expansion of the "Dirac comb"

$$\sum_{m\in\mathbf{Z}} \delta(x - 2m\pi) = \frac{1}{2\pi}\sum_{n\in\mathbf{Z}} \exp(inx)$$

3.62
(1) Write the Fourier expansion in $L^2(-\pi,\pi)$ of the function (see Problem 1.20, q.(1))

$$f(x) = \begin{cases} -1 & \text{for } -\pi < x < 0 \\ 1 & \text{for } \quad 0 < x < \pi \end{cases}$$

in terms of the orthogonal complete set $\{1, \cos nx, \sin nx , n = 1, 2, \ldots\}$ (actually, the subset $\{\sin nx\ n = 1, 2, \ldots\}$ is enough). Consider then the periodic prolongation $\widetilde{f}(x)$ to all $x \in \mathbf{R}$ of $f(x)$ with period 2π: evaluate the first derivative of $\widetilde{f}(x)$ and of its Fourier expansion (the expansion is automatically periodic) to obtain the identity

$$\frac{2}{\pi}\sum_{n\in\mathbf{Z}} \cos\big((2n+1)x\big) = \sum_{m\in\mathbf{Z}} \delta\big(x - 2m\pi\big) - \sum_{m\in\mathbf{Z}} \delta\big(x - (2m+1)\pi\big)$$

(2) Proceed as in (1) for the function (cf. Problem 1.21, q.(2))

$$f(x) = \begin{cases} x + \pi & \text{for} \quad -\pi < x < 0 \\ x - \pi & \text{for} \quad 0 < x < \pi \end{cases}$$

to obtain again the Fourier expansion of the "Dirac comb" as in the problem above (what is the derivative of $f(x)$?).

3.2.2 *Fourier Transform, Distributions and Linear Operators*

In this subsection we will consider, in the context of Fourier transform (see the Introduction to the Sect. 3.1.2) some examples of linear operators whose action is extended from the Hilbert space $L^2(\mathbf{R})$ to the linear space of distributions $\mathscr{S}'$. Also many of the problems proposed in the following subsections can be stated in terms of operators of this type; indeed, whenever there is a linear relationship between an "input" $a(t)$ and the corresponding "output" $b(t)$, one can define a linear operator T simply putting $T\big(a(t)\big) = b(t)$.

3.63

(1) *(a)* Using Fourier transform, study the convergence as $n \to \infty$ of the sequence of operators $T_n : L^2(\mathbf{R}) \to L^2(\mathbf{R})$ (already considered in Problem 3.18, q.(4))

$$T_n\, f(x) = \int_{-\infty}^{+\infty} f(x) \frac{\sin\big(n(x-y)\big)}{\pi(x-y)}\, dy$$

Recognize that this type of convergence can be viewed as the statement, in the language of the operators in Hilbert space, that the sequence of functions $g_n(x) = \sin nx/(\pi x)$ converges to $\delta(x)$ (in $\mathscr{S}'$, of course);

(b) changing the variable x into the time variable t, the operators T_n can be interpreted as ideal filters of "low frequencies" $|\omega| \le n$ (see Problem 3.18). Construct the operators S_n of the filters for "high frequencies" $|\omega| \ge n$. What is $\lim_{n\to\infty} S_n$?

(2) Consider now the family of operators studied in Problem 3.19, q.(6)

$$T_a f(x) = \int_{-\infty}^{+\infty} f(y) \frac{a}{a^2 + (x-y)^2}\, dy$$

and study their convergence as $a \to 0^+$. Verify that the same remark holds as for the sequence of operators seen in (1)(*a*). Construct similar examples of sequences (or families) of operators replacing $g_n(x)$ with other sequences (or families) of functions tending to $\delta(x)$.

3.64

It is well-known that the operator $T f(x) = x\, f(x)$ in $L^2(\mathbf{R})$ has no eigenvectors (see Problem 1.74, q.(2)). However, looking for "eigenvectors" in the vector space $\mathscr{S}'$, one easily sees that for each "eigenvalue" $\lambda \in \mathbf{R}$, there is the "eigenvector" $\delta(x-\lambda)$, indeed $(x-\lambda)\,\delta(x-\lambda) = 0$. Using this fact:

(1) Look for "eigenvalues" and "eigenvectors" of the operator

$$T f(x) = \sin x\, f(x)\,, \qquad x \in \mathbf{R}$$

(2)(*a*) Using Fourier transform, look for "eigenvalues" and "eigenvectors" of the operator

$$T f(x) = f(x-1)\,, \qquad x \in \mathbf{R}$$

Look in particular for the "eigenvectors" corresponding to the "eigenvalue" $\lambda = 1$.

(*b*) Essentially the same question as in (*a*): what is the most general form of the Fourier transform $\widehat{f}(\omega)$ of a periodic function $f(x)$ of period 1 (or period τ: $f(x) = f(x-\tau)$, $\forall x \in \mathbf{R}$)?

3.65

Consider the operator defined in $L^2(\mathbf{R})$

$$T f(x) = f(x) - f(x-1)$$

Using Fourier transform:

(1) Find $\|T\|$. Is there any $f_0(x) \in L^2(\mathbf{R})$ such that $\|T f_0\| = \|T\|\,\|f_0\|$?

(2) Are there eigenvectors of T in $L^2(\mathbf{R})$? and in $\mathscr{S}'$?

(3) Find $\mathtt{Ker}\, T$ and $\mathtt{Ran}\, T$, specifying if $\mathtt{Ran}\, T = L^2(\mathbf{R})$ or at least is dense in it.

(4) More in general, let T_{ab} be the operator

$$T_{ab} f(x) = f(x-a) - f(x-b)\,, \qquad a,\, b \in \mathbf{R}$$

study the limit of T_{ab} for $b \to a$.

3.66

(1) In a given linear system, if the input is $a(t) = \theta(t)\exp(-t)$ then the corresponding output is $b(t) = \exp(-|t|)$. Using Fourier transform, show that this uniquely defines the Green function $G(t) \in L^2(\mathbf{R})$ of the system (with the usual notations $b(t) = \big(G * a\big)(t)$).

(2) Consider the linear operator $T : L^2(\mathbf{R}) \to L^2(\mathbf{R})$ defined by

$$T\,\big(a(t)\big) = b(t) = \big(G * a\big)(t)$$

where $a(t)$ and $b(t)$ are given in (1). Find $\|T\|$ and check if there is some input $a(t) \in L^2(\mathbf{R})$ such that $b(t) = \lambda\, a(t)$.

(3) It may appear surprising that the operator T can be completely defined in $L^2(\mathbf{R})$ giving only *one* information, namely the result obtained when T is applied to the *single* function $a(t)$, according to (1). Give an explanation of this fact (see however also next problem).

3.67

(1) Differently from the case considered above, it can happen that giving a single input $a(t)$ with the corresponding output $b(t) = \big(G * a\big)(t)$ is *not* enough to define the Green function of the system. Discuss the following cases:

(*a*) $a(t) = \exp(-|t|),\ b(t) = \theta(t)\exp(-t)$

(*b*) $a(t) = \dfrac{\sin t}{t},\ b(t) = 0$

(*c*) $a(t) = \dfrac{\sin 2t}{t},\ b(t) = \dfrac{\sin t}{t}$

(*d*) $a(t) = \dfrac{\sin t}{t},\ b(t) = \dfrac{\sin 2t}{t}$

(*e*) $a(t) = \begin{cases} -\exp(t) & \text{for } t < 0 \\ \exp(-t) & \text{for } t > 0 \end{cases},\ b(t) = \exp(-|t-1|) - \exp(-|t+1|)$

(*f*) $a(t) = \begin{cases} -\exp(t) & \text{for } t < 0 \\ \exp(-t) & \text{for } t > 0 \end{cases},\ b(t) = \exp(-|t-1|) + \exp(-|t+1|)$

(*g*) $a(t) = \begin{cases} 1 & \text{for } |t| < 1 \\ 0 & \text{elsewhere} \end{cases},\ b(t) = \exp(-|t-1|) - \exp(-|t+1|)$

(*h*) $a(t) = \begin{cases} 1 & \text{for } |t| < 1 \\ 0 & \text{elsewhere} \end{cases},\ b(t) = 0$

(2) For completeness and comparison, consider also the following cases, where both $a(t)$ and $b(t) \notin L^2(\mathbf{R})$:

$$a(t) = \delta'(t),\ b(t) = \delta(t+1) \pm \delta(t-1)$$

3.68

Consider a linear system described by a Green function $G(t)$ as usual, and introduce the linear operator $T : L^2(\mathbf{R}) \to L^2(\mathbf{R})$ defined by $T\big(a(t)\big) = b(t) = \big(G * a\big)(t)$. Assume that $G(t) \in L^1(\mathbf{R}) \cap L^2(\mathbf{R})$

(1) Using Fourier transform, show that if the input $a(t) \in L^2(\mathbf{R})$ then also the output $b(t) \in L^2(\mathbf{R})$.

(2) Find $\|T\|$.

(3) Assume that some input $a(t)$ is "a good approximation", in the sense of the norm $L^2(\mathbf{R})$, of another input $\widetilde{a}(t)$ (i.e., $\|a(t) - \widetilde{a}(t)\|_{L^2(\mathbf{R})} < \varepsilon$ for some "small" $\varepsilon > 0$). Show that also the corresponding output $b(t)$ approximates $\widetilde{b}(t)$ and find some constant C such that $\|b(t) - \widetilde{b}(t)\|_{L^2(\mathbf{R})} < C\varepsilon$.

(4) Show that the properties in (1), (2) and (3) are still true under the only hypothesis that the Fourier transform $\widehat{G}(\omega)$ is a bounded function.

(5) What changes if $\widehat{G}(\omega)$ is unbounded?

3.2.3 *Applications to ODE's and Related Green Functions*

3.69

(1) Consider the differential equation

$$\dot{x} + ax = f(t)\,, \qquad x = x(t)\,,\ a > 0\,, \quad t \in \mathbf{R}$$

Find the Fourier transform $\widehat{G}(\omega)$ of the Green function of this equation and then the function $G(t)$ (see also Problem 3.4). Find *the most general* Green function $G(t)$.

(2) Obtain again the most general Green function $G(t)$ by calculation "step by step": start with the solution $G_-(t)$ obtained integrating the equation $\dot{G}_- + aG_- = 0$ for $t < 0$, and the solution $G_+(t)$ obtained integrating the equation $\dot{G}_+ + aG_+ = 0$ for $t > 0$; deduce then the "global" solution of the equation $\dot{G} + aG = \delta(t)$ imposing the suitable discontinuity condition at the point $t = 0$ (to $G_-(t)$ as $t \to 0^-$ and to $G_+(t)$ as $t \to 0^+$).

(3) The same questions (1) and (2) for the equation

$$\dot{x} - ax = f(t)\,, \qquad x = x(t)\,,\ a > 0$$

(4) Among the Green functions obtained for the equation in (1), is there a Green function which is causal and belongs to $L^2(\mathbf{R})$ (or to $\mathscr{S}'$)? And among the Green functions obtained for the equation in (3)?

3.70

Consider again the two equations

$$\dot{x} \pm ax = f(t)\,, \qquad x = x(t)\,,\ a > 0$$

(1) Let $G_\pm(t)$ be the Green function of these equations obtained *using Fourier transform* (see problem above):

(a) find their limits as $a \to 0^+$ specifying in what sense these limits exist;

(b) do these limits coincide?

(c) do these limits solve the equation $\dot{x} = \delta(t)$?

(2) Find the most general solution of the equation

$$\dot{x} = \delta(t)$$

Is there any relationship between these solutions and the limits obtained in $(1)(a)$?

3.71

Consider the equation

$$\dot{x} + x = f(t)\,, \qquad x = x(t)$$

(1) Let $f(t) = \theta(t)\exp(-\alpha t)$, $\alpha \neq 1$.

(a) Solve the equation by means of Fourier transform. Only one solution is obtained; considering the solutions of the homogeneous equation $\dot{x}_0 + x_0 = 0$, write then the most general solution of the equation, and find the solution which satisfies the condition $x(1) = 1$.

(b) The same with $\alpha = 1$.

(2) The same with $f(t) = \mathtt{sgn}\,t$.

3.72

Consider the equation

$$\dot{x} + ix = f(t)\,, \qquad x = x(t)$$

(1) Using Fourier transform find the most general Green function of this equation.

(2) Find the most general solution of the equation if $f(t) = \theta(t)\exp(-t)$, and then the solution which satisfies the condition $x(1) = 1$.

(3) The same as in (2) if $f(t) = \mathtt{sgn}\,t$.

(4) Discuss the main difference between this and the previous problem: how many Green functions are obtained in each case using Fourier transform?

3.73

Consider the equation (use Fourier transform in all questions)

$$\dot{x} + ax = f(t)\,, \qquad x = x(t)\,,\ a > 0$$

(1) Find the solution $x(t)$ if $f(t) = \sin t$.

$(2)(a)$ Let $f(t) = h(t-1) - h(t+1)$, where $h(t) \in L^2(\mathbf{R})$. Show that the corresponding solution $x = x_a(t) \in L^2(\mathbf{R})$ and find some constant C such that $\|x_a(t)\|_{L^2(\mathbf{R})} \le C\|h(t)\|_{L^2(\mathbf{R})}$.

(b) Let $\widetilde{x}(t) = \lim_{a\to 0^+} x_a(t)$: show that also $\widetilde{x}(t) \in L^2(\mathbf{R})$ and find a constant C such that $\|\widetilde{x}(t)\|_{L^2(\mathbf{R})} \le C\|h(t)\|_{L^2(\mathbf{R})}$.

(3) What changes in (2) if $f(t) = h(t-1) + h(t+1)$?

3.74

Consider the following ODE for the unknown function $x = x(t)$, where $f = f(t)$ is given; use Fourier transform:

$$a\,\dot{x} + b\,x = \dot{f}\,, \qquad a, b > 0$$

(1) Show that if $f(t) \in L^2(\mathbf{R})$ then also $x(t) \in L^2(\mathbf{R})$ and find a constant C such that $\|x\|_{L^2} \le C\,\|f\|_{L^2}$.

(2) Put $a = b = 1$. Show that the solution $x(t)$ can be written as $x(t) = f(t) + x_1(t)$ where $x_1(t)$ satisfies the ODE ...

(3) Put $a = b = 1$ and find $x(t)$ in the cases:

$$f(t) = \delta(t) \ \text{ and } \ f(t) = \theta(t)$$

3.75

The equation of the motion of a particle of mass $m = 1$ subjected to an external force $f(t)$ and to a viscous damping (cf. Problem 3.12), denoting by $v(t)$ its velocity, is

$$\dot{v} + \beta v = f(t)\,, \qquad \text{let } \ f(t) = \theta(t)\exp(-t)$$

(1) Let $\beta > 0$ and $\beta \ne 1$. Use Fourier transform to find $\widehat{v}(\omega) = \mathscr{F}\big(v(t)\big)$ and $v = v(t)$.

(2) Let now $\beta = 0$.

(a) The corresponding solution $v^{(0)}(t)$ can be found in several ways:

(i) evaluate the limit $\lim_{\beta\to 0^+}$ of the solution $v(t)$ found in (1). Does this limit exist in $L^2(\mathbf{R})$?

(ii) solve the equation $\dot{v}^{(0)} = \theta(t)\exp(-t)$ by direct integration (with no use of Fourier transform, and under the condition $v^{(0)}(t) = 0$ if $t \le 0$)

(iii) use the Fourier transform of the equation $\dot{v}^{(0)} = \theta(t)\exp(-t)$ (recall that $\lim_{\varepsilon\to 0^+} 1/(x + i\varepsilon) = \cdots$)

(b) Find now $v^{(0)}(+\infty)$ and the final kinetic energy $\big(v^{(0)}(+\infty)\big)^2/2$ of the particle.

(3)(a) Again, use three different procedures to find the total work W_f done by the

force (with generic $\beta > 0$) and its limit $W_f^{(0)}$ when $\beta \to 0^+$:
(i) evaluate the integral

$$W_f = \int_L f\,dx = \int_{-\infty}^{+\infty} f(t)v(t)\,dt$$

(where L is the space covered by the particle) using the given $f(t)$ and the expression of $v(t)$ obtained in (1) and then find its limit

$$W_f^{(0)} = \lim_{\beta\to 0^+} W_f$$

Is this limit equal to $\displaystyle\int_{-\infty}^{+\infty} f(t)v^{(0)}(t)\,dt$? (why?).
(ii) use now Parseval identity: write $W_f = \big(f, v\big)$ as an integral in the variable ω in terms of the expressions of $\widehat{f}(\omega)$ and $\widehat{v}(\omega)$ obtained above, evaluate explicitly this integral by integration in the complex plane ω, and then find its limit as $\lim_{\beta\to 0^+}$
(iii) as a third possibility for finding the limit $W_f^{(0)}$, write W_f as an integral in the variable ω as in *(ii)* and, *before* evaluating this integral, perform first the $\lim\limits_{\beta\to 0^+}$ etc.
(These various ways for obtaining the same result propose some different relevant aspects and provide useful exercises!)
(b) Verify that if $\beta = 0$ the work $W_f^{(0)}$ is (expectedly!) equal to the final kinetic energy of the particle, found in (2)*(b)* .

3.76
Consider now the same problem as the above one but extending to the case of a *general* force $f(t) \in L^1(\mathbf{R}) \cap L^2(\mathbf{R})$ (and $f(t)$ real, of course).
(1) Let $\beta > 0$; write the Fourier transform $\widehat{v}(\omega) = \mathcal{F}\big(v(t)\big)$ in terms of the Fourier transform of the Green function $\widehat{G}(\omega)$ of the equation and of the applied force $\widehat{f}(\omega) = \mathcal{F}\big(f(t)\big)$.
(2) Write the work W_f done by the force (cf. Problems 3.12–3.13 and the previous one)

$$W_f = \int_L f\,dx = \int_{-\infty}^{+\infty} f(t)v(t)\,dt = \big(f, v\big)$$

(where L is the space covered by the particle) using Parseval relation as an integral in the variable ω in terms of $\widehat{f}(\omega)$ and $\widehat{G}(\omega)$. Evaluate then the limit $W_f^{(0)} = \lim\limits_{\beta\to 0^+} W_f$.
Hint: recall that $f(t)$ is real, therefore $\widehat{f^*}(\omega) = \cdots$, and that $\lim\limits_{\varepsilon\to 0^+} 1/(x + i\varepsilon) \to \cdots$.
(3) When $\beta = 0$, the equation $\dot{v} = f(t)$ can be integrated directly (with no use of Fourier transform): show that $v^{(0)}(+\infty) = \widehat{f}(0)$ (under the condition $v(-\infty) = 0$), and that, as expected, $W_f^{(0)}$ is equal to the final kinetic energy of the particle.

(4) When $\beta = 0$, under what condition on the applied force $f(t)$, does the work $W_f^{(0)}$ done by $f(t)$ vanish?

3.77
Consider the following equation and use Fourier transform

$$\dot{x} + iax = \delta(t) - \delta(t-1), \qquad x = x(t), a \in \mathbf{R}$$

(1) For what values of $a \in \mathbf{R}$ are there solutions $x(t) \in L^2(\mathbf{R})$?
(2) Find the most general solution of this equation in the cases $a = 0,\ a = \pi,\ a = 2\pi$.
(3) The same questions (1) and (2) for the equation

$$\dot{x} + iax = \delta(t) + \delta(t-1)$$

3.78
(1) Using Fourier transform, find the Green function(s) of the ODE

$$\ddot{x} + x = f(t), \qquad x = x(t)$$

How many Green functions are obtained using Fourier transform? and does one obtain in this case the most general Green function? Write in particular the causal Green function: does it belong to $L^2(\mathbf{R})$? to $\mathcal{S}'$? (see also Problem 3.56 *(d)*).

(2) Find again the Green functions proceeding by direct calculation "step by step", as in Problem 3.69: here, one has to impose the continuity condition at $t = 0$ to $G_\pm(t)$, and a suitable discontinuity condition to $\dot{G}_\pm(t)$ ....

3.79
The same questions (1) and (2) as in the above problem for the equation

$$\ddot{x} = f(t), \qquad x = x(t)$$

(see also Problem 3.54 *(d)*).

3.80
Using Fourier transform, find the Green function of each one of the following equations

$$\ddot{x} - x = f(t) \ ; \ \ddot{x} \pm \dot{x} = f(t), \qquad x = x(t)$$

How many Green functions are obtained proceeding through Fourier transform? Explain why one does not obtain the expected ∞^2 solutions. Find then the most general Green functions (see also Problem 3.56 *(h)*, *(f)*).

3.81

(1) By means of Fourier transform, find the most general solution of the following equation and find in particular the solution which "respects causality"[2]

$$\ddot{x} = \theta(t)\exp(-t), \qquad x = x(t),\ t \in \mathbf{R}$$

Hint: cf. Problem 3.54 *(d)* and use a decomposition as $\dfrac{1}{y^2(y-c)} = \dfrac{a_1}{y} + \dfrac{a_2}{y^2} + \dfrac{b}{y-c}$.

(2) Obtain again the general solution by direct repeated integration (without using Fourier transform; impose continuity at $x = 0$ of $x(t)$ and $\dot{x}(t)$)

3.82

Solve by means of Fourier transform the equation

$$\ddot{x} - x = \theta(t)\exp(-2t), \qquad x = x(t),\ t \in \mathbf{R}$$

How many solutions are obtained in this way (cf. Problem 3.56 *(h)*)? Write then the most general solution. Find in particular the solution respecting causality (see previous problem). Is this solution in $L^2(\mathbf{R})$? in $\mathscr{S}'$? Is there a solution in $L^2(\mathbf{R})$?

3.83

The same questions as in the above problem for the equation (cf. Problem 3.56 *(f)*)

$$\ddot{x} + \dot{x} = \theta(t)\exp(-t), \qquad x = x(t),\ t \in \mathbf{R}$$

3.84

Solve by means of Fourier transform the equation (cf. Problem 3.56 *(d)*)

$$\ddot{x} + x = \theta(t)\exp(-t), \qquad x = x(t),\ t \in \mathbf{R}$$

Does one obtain in this way the most general solution? Find in particular the solution respecting causality. Is this solution in $L^2(\mathbf{R})$? in $\mathscr{S}'$?

3.85

(1) *(a)* Find the most general solution $x(t)$ of the equation

$$\dot{x} + i x = \exp(-i\alpha t), \qquad t \in \mathbf{R},\ \alpha \neq 1$$

both by means of elementary integration rules for linear ODE's, and by means of Fourier transform.

(b) The same questions in the "resonant case" $\alpha = 1$, i.e.,

---

[2] Assuming that the "input" $f(t) = 0$ for $t < t_0$, the solution respecting causality is requested to be zero for $t < t_0$ (in the present case $t_0 = 0$ and $x(t) = 0$ for $t < 0$; see the Introduction to this chapter).

$$\dot{x} + i\,x = \exp(-it)$$

Hint: recall that $(y-1)\delta'(y-1) = \cdots$

(2) The same as in (1) for the equation

$$\ddot{x} + x = \sin\alpha t$$

with $\alpha \neq 1$ and $\alpha = 1$. *Hint:* in the resonant case $\alpha = 1$, show first that

$$(y^2-1)\big(\delta'(y-1) \pm \delta'(y+1)\big) = A\big(\delta(y-1) \mp \delta(y+1)\big) \quad \text{where } A = \cdots$$

3.86

Let $f_n(t)$ be the functions

$$f_n(t) = \begin{cases} \sin t & \text{for} \quad |t| \le 2n\pi \\ 0 & \text{for} \quad |t| \ge 2n\pi \end{cases}, \qquad n = 1, 2, \ldots, \; t \in \mathbf{R}$$

(1) *(a)* Calculate the second derivative $\ddot{f}_n(t)$

(b) Without using Fourier transform but using the result in *(a)*, find the most general solution of the equation

$$\ddot{x} + x = \delta(t + 2n\pi) - \delta(t - 2n\pi)$$

(2) Let

$$\widehat{g}_n(\omega) = \frac{\exp(-2in\pi\omega) - \exp(+2in\pi\omega)}{1 - \omega^2}$$

Show that $\widehat{g}_n(\omega) \in L^2(\mathbf{R})$; find then $g_n(t) = \mathscr{F}^{-1}\big(\widehat{g}_n(\omega)\big)$.
Hint: notice that $(-\omega^2 + 1)\widehat{g}_n(\omega) = \cdots$ and therefore $g_n(t)$ satisfies the differential equation …; recall that $\widehat{g}_n(\omega) \in L^2(\mathbf{R})$.

(3) Find the limits (in $\mathscr{S}'$, of course)

$$\lim_{n\to+\infty}\big(\exp(-2in\pi\omega) - \exp(+2in\pi\omega)\big) \quad \text{and} \quad \lim_{n\to+\infty}\frac{\exp(-2in\pi\omega) - \exp(+2in\pi\omega)}{1 - \omega^2}$$

In all the above problems of this subsection, the independent variable has been the time t, and in most cases the notion of causality has been introduced. In the following problems of this subsection, we will introduce as independent variable the "position" $x \in \mathbf{R}$, and the unknown variable will be denoted by $u = u(x)$. The procedure for solving differential equations is clearly exactly *the same as before, but the physical interpretation changes. For instance, there no reason to impose a condition similar to the causality: indeed, assuming that there is some point x_0 such that the applied term $f(x)$ is zero for all $x < x_0$, one expects that the effect propagates also to the points "at the left" of x_0, i.e. also to the points $x < x_0$. Similarly, there is no reason to impose to the Green function to be zero for $x < 0$. So, instead of causality conditions, one may have to impose different conditions depending on the physical situations (e.g., vanishing of the solution at $x \to +\infty$, some boundary conditions, suitable continuity or discontinuity properties, etc.).*

3.87
Consider the equation

$$-u''(x) = \delta(x - x_1), \qquad 0 \le x \le 1$$

where x_1 is any fixed point with $0 < x_1 < 1$, and with boundary conditions

$$u(0) = u(1) = 0$$

This problem amounts, for instance, to looking for *stationary* (i.e., time-independent) solutions $u = u(x)$ of the d'Alembert equation describing an elastic string with fixed end points in $x = 0$ and $x = 1$, in the presence of an applied force $f(x)$

$$u_{tt} - u_{xx} = f(x)$$

when the force is "concentrated" at the point x_1. In other words, this amounts to finding the Green function of this problem: clearly, the Green function will depend also on the point x_1, and one expects to have $u = u(x, x_1)$, not $u = u(x - x_1)$, because the boundary conditions "destroy" any space translational invariance, differently from the case of many (previously considered) problems depending on time, where the presence of time invariance produces Green function of the form $G = G(t - t')$ (cf. the Introduction of this chapter).

Use three different procedures:

(1) Apply Fourier transform to the equation $-u'' = \delta(x - x_1)$, assuming here $x \in \mathbf{R}$, find $\widehat{u}(k) = \mathscr{F}\big(u(x)\big)$ (see Problem 3.54 *(d)*) and evaluate the inverse Fourier transform $u(x)$; impose then to this $u(x)$ the boundary conditions $u(0) = u(1) = 0$.

(2) Solve the equation $-u'' = \delta(x - x_1)$ by calculation "step by step", as in Problem 3.69: find the solution $u_-(x)$ of the equation $u''_- = 0$ for $0 < x < x_1$, and the solution $u_+(x)$ of the equation $u''_+ = 0$ for $x_1 < x < 1$; impose the boundary conditions $u_-(0) = 0$ and $u_+(1) = 0$, impose finally the continuity condition at $x = x_1$ to $u_\pm(x)$ and the suitable discontinuity condition at $x = x_1$ to $u'_\pm(x)$.

(3) Solve the equation $-u'' = \delta(x - x_1)$ by direct integration (without using Fourier transform, recall that $\frac{d}{dx}\theta(x - x_1) = \cdots$), etc.

3.88

By means of Fourier transform, find the most general solution of the equation (cf. Problem 3.81)

$$u_{xx} = \theta(x)\exp(-x)\,, \qquad u = u(x),\ x \in \mathbf{R}$$

Find in particular the solution vanishing when $x \to +\infty$, and the solution satisfying the boundary conditions $u(-1) = u(1) = 0$.

3.89

Solve by means of Fourier transform the equation (cf. Problem 3.84)

$$u_{xx} + u = \theta(x)\exp(-x)\,, \qquad u = u(x),\ x \in \mathbf{R}$$

Find in particular the solution satisfying the boundary conditions $u(0) = u(\pi/2) = 0$. Verify that it is impossible to have a solution satisfying the boundary conditions $u(0) = u(\pi) = 0$: why?

3.2.4 *Applications to General Linear Systems and Green Functions*

3.90

(1) Consider any linear system defined by a Green function $G(t)$ where input $a(t)$ and output $b(t)$ are related by $b(t) = \big(G * a\big)(t)$ as usual. Let the input be a monochromatic wave $a(t) = \exp(-i\omega_0 t)\,, \forall t \in \mathbf{R}$. Using Fourier transform, show that also the output (if not zero) is a monochromatic wave differing from the input in presenting an "amplification" and a "phase shift".

(2) Find the output $b(t)$ if the input is the superposition of two monochromatic waves as $a(t) = \exp(-i\omega_1 t) + \exp(-i\omega_2 t)\,, \forall t \in \mathbf{R}$, with $\omega_1 \neq \omega_2$.

3.91

(1) The Fourier transform of the Green function $G = G_n(t)$ of a linear system is given by

$$\widehat{G}_n(\omega) = \frac{1}{(\omega - i)^n}\,, \qquad n = 1, 2, \ldots$$

Without calculating $G_n(t)$, specify for what n the Green function $G_n(t)$ is a *real* function.

(2) Let now $n = 2$ in the above Green function and let $a(t) = \sin t$ be the input applied to the system. Find $\widehat{a}(\omega)$ and the corresponding output $b(t)$. Is the output a real function? (cf. (1)).

(3) The same as in (2) if $a(t) = \sin|t| = \mathtt{sgn}\,t\ \sin t$.

3.92
Consider a linear system described by a Green function $G(t)$ as usual.
(1)(*a*) Let the Fourier transform $\widehat{G} = \widehat{G}_0(\omega)$ of the Green function be given by

$$\widehat{G}_0(\omega) = -i\omega$$

What is the relationship between the input $a(t)$ and the output $b(t)$? Find $b(t)$ if $a(t) = \theta(t)$.

(*b*) Using Fourier transform, show the identity $T * \delta^{(n)} = D^{(n)}T$, and verify that the question (*a*) is a particular case of this.

(2) Consider now the following "approximation" $\widehat{G}_\varepsilon(\omega) \in L^2(\mathbf{R})$ of $\widehat{G}_0(\omega)$

$$\widehat{G}_\varepsilon(\omega) = \frac{-i\omega}{(1 - i\varepsilon\omega)^2}, \qquad \varepsilon > 0$$

(verify that indeed $\widehat{G}_\varepsilon(\omega) \to \widehat{G}_0(\omega)$ in $\mathscr{S}'$). Let $a(t) = \theta(t)$. Find the corresponding output $b_\varepsilon(t)$ if the Green function is given by $\widehat{G}_\varepsilon(\omega)$ and show that indeed $b_\varepsilon(t)$ is an approximation of ....

3.93
A linear system is defined by a Green function whose Fourier transform is

$$\widehat{G}_\tau(\omega) = \frac{\omega^2}{(1+\omega^2)^2} \exp(i\omega\tau), \qquad \tau \in \mathbf{R}$$

(1) If the input is

$$a(t) = \begin{cases} 1 & \text{for } t < 0 \\ 2 & \text{for } t > 0 \end{cases}$$

is it possible to say that the corresponding output $b = b_\tau(t)$ belongs to $L^2(\mathbf{R})$ (without calculating it)?

(2) What properties (continuity, differentiability, behavior as $|t| \to \infty$) can be expected for $b_\tau(t)$?

(3) Study the convergence as $\tau \to \infty$ of $b_\tau(t)$.

3.94
Consider a linear system defined by the Green function

$$G = G_T(t) = \begin{cases} 1 & \text{for } 0 < t < T \\ 0 & \text{elsewhere} \end{cases}, \qquad T > 0$$

with input $a(t)$ and output $b = b_T(t)$ related by the usual rule $b = G * a$.

(1) Using Fourier transform find $b = b_T(t)$ if $a(t) = t\ \exp(-t^2)$

(2) Find $\lim_{T\to\infty} b_T(t)$. Does this limit exist in the norm $L^2(\mathbf{R})$? in the sense of $\mathcal{S}'$? Confirm the conclusion examining $\lim_{T\to\infty} \widehat{b_T}(\omega)$.

(3) Let now $G(t) = \theta(t)$: using Fourier transform find $b(t)$ if $a(t) = t\ \exp(-t^2)$; does this $b(t)$ coincide with the limit obtained in (2)?

3.95

Consider a linear system defined by the Green function (use Fourier transform)

$$G = G_c(t) = \frac{\sin ct}{\pi t}, \qquad c > 0$$

with input $a(t)$ and output $b = b_c(t)$ related by the usual rule $b_c = G_c * a$.

(1) Is there some input $a(t) \neq 0$ such that $b_c(t) = 0$ for all $t \in \mathbf{R}$?

(2) Let $a(t) \in L^2(\mathbf{R})$: what properties of the output $b_c(t)$ can be expected? (specify if $b_c(t) \in L^1(\mathbf{R}) \cap L^2(\mathbf{R})$, discuss its continuity and differentiability, boundedness, behavior at $|t| \to \infty$).

(3) Find $\lim_{c\to\infty} G_c(t)$. If $a(t) \in L^2(\mathbf{R})$, find $\lim_{c\to\infty} b_c(t)$; does this limit exist also in the sense of the norm $L^2(\mathbf{R})$?

3.96

Consider a linear system defined by the Green function

$$G = G_T(t) = \begin{cases} 1 & \text{for } |t| < T \\ 0 & \text{elsewhere} \end{cases}, \qquad T > 0$$

with input $a = a(t)$ and output $b = b_T(t)$ related by the usual rule $b_T = G_T * a$.

(1) Is there some nonzero input $a(t) \in L^2(\mathbf{R})$ such that $b_T(t) = 0$ for all $t \in \mathbf{R}$? and if $a(t) \in \mathcal{S}'$?

(2) Let $a(t) \in L^2(\mathbf{R})$: what properties of the output $b_T(t)$ can be expected? (specify if $b_T(t) \in L^1(\mathbf{R}) \cap L^2(\mathbf{R})$, discuss its continuity and differentiability, boundedness, behavior at $|t| \to \infty$).

(3) Find $\lim_{T\to+\infty} G_T(t)$ and $\lim_{T\to+\infty} \widehat{G}_T(\omega)$.

(4) Let $a(t) \in L^1(\mathbf{R}) \cap L^2(\mathbf{R})$: find $\lim_{T\to+\infty} \widehat{b_T}(\omega)$ and $\lim_{T\to+\infty} b_T(t)$.

3.97

Assume that the input $a(t)$ and the corresponding output $b(t)$ of a system are related by the rule

$$a(t) = b(t) + \alpha \int_{-\infty}^{t} \exp\big(-(t-t')\big) b(t')\, dt'$$

(1) Using Fourier transform, find the Green function $G(t)$, defined as usual as $b = G * a$, in each one of the following cases:

$$\alpha = 1 \ ; \quad \alpha = -1 \ ; \quad \alpha = -2$$

(2) Verify (*without using in this question Fourier transform*) that the given equation with $a(t) = 0$ and $\alpha = -2$ admits a solution of the form $b_0(t) \propto \exp(\beta t)$, with $\beta \in \mathbf{R}$ to be determined.

(3) Assuming that the only solution of the equation with $\alpha = -2$ in the "homogeneous" case $a(t) = 0$ is that found in (2), write the most general Green function of the equation in the case $\alpha = -2$ and specify if there is a *causal* Green function.

3.98

The input $a(t)$ and the output $b(t)$ of a linear system are related by the rule

$$\dot{b}(t) = a(t) - a(t-1)$$

(1) Using Fourier transform find the most general Green function $G(t)$ of this problem.

(2) Both by direct integration (without using Fourier transform), and using Fourier transform, find $b(t)$ in the cases

$$a(t) = t \exp(-t^2) \quad \text{and} \quad a(t) = \theta(t)$$

(3) The same questions (1) and (2) if the equation is

$$\dot{b}(t) = a(t) + a(t-1)$$

3.99

The input $a(t)$ and the output $b(t)$ of a linear system are related by the rule

$$\ddot{b}(t) = 2a(t) - a(t-1) - a(t+1)$$

(1) Using Fourier transform, find the most general Green function $G(t)$ of the problem (see Problems 3.29, q.(2) and 3.54 *(h)*) and specify if there is a Green function $G_0(t)$ belonging to $L^2(\mathbf{R})$.

(2) Find the most general solution $b(t)$ in the cases

$$a(t) = \dot{\delta}(t) \, ; \quad a(t) = t \, ; \quad a(t) = \theta(t) \exp(-t)$$

3.100

(1) Assume that in a linear system the output $b(t)$ is related to the input $a(t)$ by the rule, where $G(t)$ is given

$$b(t) = \int_{-\infty}^{+\infty} G(t+t')\, a(t')\, dt'$$

(*a*) Find $b(t)$ in the special case $G(t+t') = \delta(t+t')$.

(*b*) Clearly, the above relation between $a(t)$ and $b(t)$ cannot be expressed in the form $b = G * a$ and, as a consequence, also $\widehat{b}(\omega) = \widehat{G}(\omega)\,\widehat{a}(\omega)$ is no longer true. However, introducing the change of variable $t'' = -t'$, the above relation can be written as $b = G * a^{(-)}$ where $a^{(-)} = \cdots$. Using this trick, evaluate $b(t)$ using Fourier transform in the case

$$G(t) = \theta(t)\exp(-t) \;\text{ and }\; a(t) = \theta(t)\exp(-\alpha t)\,, \qquad \alpha > 0$$

(2) As another example of a linear system where the Green function is not of the form $G(t-t')$, consider the "multiplicative" case, where the output $b(t)$ is simply given by $b(t) = M(t)\,a(t)$. Show that also this case can be written in the general form $b(t) = \int_{-\infty}^{+\infty} G(t,t')a(t')\,dt'$, putting $G(t,t') = \cdots$.

3.101

(1) Let $g(\omega) \in L^2(\mathbf{R})$ be a function which satisfies the following property

$$\mathrm{P}\int_{-\infty}^{+\infty} \frac{g(y)}{y-\omega}\, dy \equiv -\mathrm{P}\frac{1}{\omega} * g = \pi i\, g(\omega)\,, \qquad \omega \in \mathbf{R}$$

Taking the inverse Fourier transform of this relation (in our notations, $\mathscr{F}^{-1}(g_1 * g_2) = 2\pi\, \mathscr{F}^{-1}(g_1)\mathscr{F}^{-1}(g_2)$), show that the support of the inverse Fourier transform $f(t) = \mathscr{F}^{-1}\big(g(\omega)\big)$ is $t \ge 0$. A first conclusion: the above relation is a test of causality, i.e., if the Fourier transform $g = \widehat{G}(\omega)$ of a Green function satisfies the above relation, then the Green function is causal.

(2) Choose now $g(y) = \frac{\exp(\pm i y)}{y \pm i}$ and, by integration in the complex plane, calculate the four integrals

$$\mathrm{P}\int_{-\infty}^{+\infty} \frac{\exp(\pm i y)}{(y \pm i)(y-\omega)}\, dy\,, \qquad \omega \in \mathbf{R}$$

What among the four functions $g(y)$ satisfies the property given in (1)?

(3) Confirm the result seen in (2): find the inverse Fourier transforms of the four functions $g(y)$; what of these has support in $t \ge 0$? This is an example of the second result: causality is connected to analyticity and "good behavior" (i.e., rapid vanishing) of the Fourier transform of the Green function in the *upper* half complex plane ω. Verify that indeed only one of the functions $g(y)$ satisfies both these conditions.

(4) Assume that the Fourier transform $g = \widehat{G}(\omega)$ of a Green function $G(t) \in L^2$ satisfies the property stated by the relation in (1) and therefore is causal. Separating real and imaginary parts $\widehat{G}(\omega) = \widehat{G}_1(\omega) + i\,\widehat{G}_2(\omega)$, and taking the real part of this relation deduce

$$\mathrm{P}\int_{-\infty}^{+\infty} \frac{\widehat{G}_1(y)}{y-\omega}\,dy \;=\; -\pi\,\widehat{G}_2(\omega)$$

and a similar relation taking the imaginary part. This shows that real and imaginary parts of $\widehat{G}(\omega)$ are not independent, but that the imaginary part is completely determined by the real part, and conversely. Let for instance

$$\widehat{G}_1(\omega) \;=\; \frac{1}{1+\omega^2}$$

using the above equation, deduce $\widehat{G}_2(\omega)$ and $G(t)$. As another consequence, writing $\widehat{G}(\omega) = A(\omega)\,\exp\big(i\Phi(\omega)\big)$, one concludes that the modulus $A(\omega)$ and the phase $\Phi(\omega)$ (which admit the physical meaning resp. as "amplification" and "phase distortion") are reciprocally connected. The two relations connecting $G_1(\omega)$ with $G_2(\omega)$, together with the above results, are special cases of the Kramers–Kronig relations, or—more in general—of the dispersion rules.

3.2.5 Applications to PDE's

3.102

Introduce and use in the following the Fourier transform $\widehat{u} = \widehat{u}(k,t)$ with respect to the variable x of the function $u = u(x,t)$, defined as

$$\widehat{u}(k,t) \;=\; \int_{-\infty}^{+\infty} u(x,t)\,\exp(ikx)\,dx$$

(1) Show that the heat equation (also called diffusion equation)

$$\frac{\partial^2 u}{\partial x^2} = \frac{\partial u}{\partial t}, \qquad u = u(x,t)\,;\; x \in \mathbf{R},\; t \geq 0$$

is transformed into an ODE for $\widehat{u}(k,t)$ and find its most general solution.

(2) Show that the solution $u(x,t)$ of the heat equation with the initial condition $u(x,0) = f(x)$ may be written in the form

$$u(x,t) \;=\; f(x) * G(x,t)$$

obtain the Fourier transform $\widehat{G}(k,t)$ of the Green function and the function $G(x,t)$.

(3) Find $\lim_{t\to 0^+} G(x,t)$. Explain why this is the result to be expected.

(4) Is there any $u(x,0) = f(x) \in L^2(\mathbf{R})$ (or $\in \mathcal{S}'$) such that $u(x,t=1) = e^{-1}u(x,0)$?

3.103
Consider the heat equation and use the Fourier transform and the same notations as in the previous problem. Assume that the initial condition $u(x, 0) \in L^2(\mathbf{R})$, and let T_t be the time-evolution operator

$$T_t \ : \ u(x, 0) \to u(x, t)$$

(1) Find $\|T_t\|$

(2) Study the limit as $t \to +\infty$ of the operator T_t.

(3) Study the limit as $t \to 0^+$ of the operator T_t.

3.104
Use the same notations for the heat equation as before.

(1) Let

$$u(x, 0) = f(x) = \frac{\sin x}{x}, \qquad x \in \mathbf{R}$$

(then $\widehat{f}(k) = \cdots$). For fixed $t > 0$, is it possible to establish, without calculating the corresponding solution $u(x, t)$, if

(a) $u(x, t) \in L^2(\mathbf{R})$?

(b) $\lim_{x\to\pm\infty} u(x, t) = 0$, and $u(x, t)$ is rapidly vanishing as $x \to \pm\infty$?

(c) $u(x, t) \in L^1(\mathbf{R})$?

(2) Show that if the initial condition $u(x, 0) = f(x) \in L^2(\mathbf{R})$, then for any $t > 0$ the solution $u(x, t) \in L^2(\mathbf{R})$ and is infinitely differentiable with respect to x and to t. Show that the same is also true if $f(x)$ is a combination of delta functions $\delta(x - a)$ (for any $a \in \mathbf{R}$) and derivatives thereof.

(3) Find the solution $u(x, t)$ in these cases

$$f(x) = 1 \ ; \quad f(x) = x \ ; \quad f(x) = x^2$$

(Calculations by means of Fourier transform need some care with the coefficients $2\pi, \pm i$ etc., but the results are disappointingly obvious ...)

3.105
Consider the equation

$$\frac{\partial u}{\partial t} = \frac{\partial^2 u}{\partial x^2} - a\frac{\partial u}{\partial x}, \qquad u = u(x, t),\ a > 0$$

which can be viewed as a "perturbation" (for "small" a) of the heat equation. Let $f(x) = u(x, 0)$ be the given initial condition. Use Fourier transform as in the previous problems.

(1) Show that the solution $u(x, t)$ can be written in the form $u(x, t) = f(x) * G(x, t)$ and find $G(x, t)$.

(2) Denote by $u_a(x,t)$ the solution of the given equation and resp. by $u_0(x,t)$ the solution of the heat equation with the same initial condition $f(x)$. Let $f(x) \in L^2(\mathbf{R})$: show that, for fixed t (put $t = 1$, for simplicity), $u_a(x,1)$ and $u_0(x,1)$ remain "near" in the $L^2(\mathbf{R})$ norm, i.e. that $\|u_a(x,1) - u_0(x,1)\|_{L^2(\mathbf{R})} \to 0$ as $a \to 0$. *Hint*: use Lebesgue theorem or the elementary property (especially useful for "small" y)

$$|\exp(iy) - 1| = |y|\left|\frac{\exp(iy) - 1}{y}\right| \le |y|\,, \qquad \forall y \in \mathbf{R}$$

(3) Is the same true as in (2) if $f(x) = \delta''(x)$?

(4) Show that if $f(x) = \mathtt{sgn}\,x$ then neither $u_a(x,1)$ nor $u_0(x,1)$ belong to $L^2(\mathbf{R})$, however verify that $u_a(x,1) - u_0(x,1) \in L^2(\mathbf{R})$ and the property seen in (2) still holds.

3.106

Use the same Fourier transform $\widehat{u}(k,t)$ as in the above problems but now for the d'Alembert equation

$$\frac{\partial^2 u}{\partial x^2} = \frac{\partial^2 u}{\partial t^2}\,, \qquad u = u(x,t)\,;\; x,\, t \in \mathbf{R}$$

(1) Show that the d'Alembert equation is transformed into an ODE for $\widehat{u}(k,t)$ and find its most general solution.

(2) Let the initial conditions be given by

$$u(x,0) = f(x)\quad,\quad u_t(x,0) = 0$$

show that the solution $u(x,t)$ is a superposition of two waves.

(3) Let $u(x,0) = f(x) \in L^2(\mathbf{R})$ and $u_t(x,0) = 0$: study the limit as $t \to +\infty$ of the above solution $u(x,t)$.

(4) Let the initial conditions be as before: study the limit as $t \to 0^+$ of the above solution $u(x,t)$.

3.107

Proceeding by means of Fourier transform as in the above problem for the d'Alembert equation, consider now the initial conditions

$$u(x,0) = 0\quad,\quad u_t(x,0) = g(x)$$

(1) Show that the solution $u(x,t)$ can be written in the form

$$u(x,t) = g(x) * G(x,t)$$

and find $\widehat{G}(k,t) = \mathscr{F}\big(G(x,t)\big)$ and $G(x,t)$.

(2) Find $G_t(x,t) = \partial G(x,t)/\partial t$; verify that $G_t(x,0) = \delta(x)$ and explain why this result should be expected.

(3) Let $g(x) = \theta(-x)\exp(x)$: write the Fourier transform $\widehat{u}(k,t)$ of the corresponding solution; without evaluating $u(x,t)$, but only recalling the statement of Jordan lemma, show that $u(x,t) = 0$ for $x > t > 0$ (this also should be expected: why?)

3.108

Consider the non-homogeneous d'Alembert equation

$$\frac{\partial^2 u}{\partial x^2} - \frac{1}{c^2}\frac{\partial^2 u}{\partial t^2} = \delta(x - vt)\,, \qquad u = u(x,t),\ x,t \in \mathbf{R}$$

describing (e.g.) an infinite elastic string subjected to a "delta" force traveling with velocity v.

(1) Introducing the Fourier transform $\widehat{u} = \widehat{u}(k,t)$ find the most general solution in the case $v \neq c$.

(2) The same in the case $v = c$.

3.109

Consider the PDE

$$\frac{\partial^2 u}{\partial x^2} + 2\frac{\partial^2 u}{\partial x \partial t} + \frac{\partial^2 u}{\partial t^2} = 0\,, \qquad u = u(x,t)\,;\ x \in \mathbf{R},\ t \in \mathbf{R}$$

with initial conditions

$$u(x,0) = f(x)\,,\ u_t(x,0) = g(x)$$

(1) Introducing the Fourier transform $\widehat{u}(k,t)$ as in previous problems, transform the given PDE into an ODE for $\widehat{u}(k,t)$ and find its solution in terms of the Fourier transforms $\widehat{f}(k)$ and $\widehat{g}(k)$ of the given initial conditions.

(2) If $f(x),\ g(x) \in L^2(\mathbf{R})$, is the same true, in general, for the solution $u(x,t)$ for any fixed $t \in \mathbf{R}$?

(3) Find the solution $u(x,t) = \mathscr{F}^{-1}\big(\widehat{u}(k,t)\big)$ (see q.(1)) of the PDE in terms of the given initial conditions $f(x)$ and $g(x)$.

(4) Find the solution $u(x,t)$ of the given PDE in the case

$$f(x) = \theta(x)\exp(-x)\,,\ g(x) = 0$$

3.110

Consider the Laplace equation

$$\Delta_2 u \equiv \frac{\partial^2 u}{\partial x^2} + \frac{\partial^2 u}{\partial y^2} = 0\,, \qquad u = u(x,y)$$

on the half plane $y \geq 0$, with the boundary condition $u(x,0) = f(x),\ x \in \mathbf{R}$.

Introduce and use the Fourier transform $\widehat{u} = \widehat{u}(k,y)$ with respect to the variable x of the function $u = u(x,y)$, defined as

$$\widehat{u}(k,y) = \int_{-\infty}^{+\infty} u(x,y)\exp(ikx)\,dx$$

(1) Show that the Laplace equation is transformed into an ODE for $\widehat{u}(k,y)$, whose most general solution is (it is clear that in all these problems we are considering only solutions $\in \mathscr{S}'$)

$$\widehat{u}(k,y) = A(k)\exp(ky) + B(k)\exp(-ky)$$

which can be more conveniently written

$$\widehat{u}(k,y) = A'(k)\exp(-|k|y) + B'(k)\exp(|k|y)$$

(2) Show that, imposing a boundedness condition (therefore $B'(k) = 0$, see however next problem q.(3) and Problem 3.112) and the given boundary condition, the solution can be written as a convolution product

$$u(x,y) = f(x) * G(x,y)$$

and give the expression of the Green function $G(x,y)$.

(3) Either from the expression of $\widehat{G}(k,y)$ or of $G(x,y)$, find $\lim_{y\to 0} G(x,y)$ (which is just the expected result: why?)

3.111

(1) With the same notations and assumptions as in the problem above, show that if $u(x,0) = f(x) \in L^2(\mathbf{R})$, then for $y > 0$ the solution $u(x,y)$ of the Laplace equation is infinitely differentiable with respect to x and to y. This should be expected: way?

(2) Let $f(x) = 1/(1+x^2)$. Find $\widehat{u}(k,y)$ and then $u(x,y)$. Compare with Problem 2.45, where the same result is obtained by means of a completely different procedure.

(3) Find $\widehat{u}(k,y)$ and then $u(x,y)$ if $f(x) = \sin x$ and if $f(x) = x$.

3.112

The (non-)uniqueness of the solutions $u(x,y)$ of the Laplace equation in the half plane $y \geq 0$ with given boundary condition (see previous problems) has been discussed in Problem 2.48. The non-uniqueness is due to the presence of nonzero solutions $u_0(x,y)$ of the equation $\Delta_2 u_0 = 0$ with *vanishing* boundary condition: indeed, if $u(x,y)$ solves the problem with a given non-zero boundary condition, then also $u(x,y) + c\,u_0(x,y)$ clearly solves the same problem, for any constant c. This can be reconsidered by means of Fourier transform.

(1) With the same procedure as in the above problems, show that the Fourier transform of the most general solution $u_0(x, y)$ of the Laplace equation on the half plane $y \geq 0$ with vanishing boundary condition $u_0(x, 0) = 0$ has the form

$$\widehat{u_0}(k, y) = C(k)\big(\exp(ky) - \exp(-ky)\big)$$

(2) Observe that the above $\widehat{u_0}(k, y)$ can belong to $\mathscr{S}'$ *only if* $C(k)$ has support in the single point $k = 0$, which implies that $C(k)$ must be proportional to $\delta(k)$ or to derivatives of $\delta(k)$; find then some solutions $u(x, y)$: choose e.g.

$$C(k) = \delta'(k),\ \delta''(k),\ \text{etc.}$$

3.113

(1) Find the 3-dimensional Fourier transform $\widehat{f}(k_1, k_2, k_3)$ of $f(x, y, z) = y \exp(-|x|)$. *Warning*: in 2 or more dimensions, one has $\mathscr{F}\big(f_1(x)f_2(y)\big) = \widehat{f_1}(k_1)\widehat{f_2}(k_2)$ with clear notations, therefore, e.g., $\mathscr{F}\big(f(x)\big) = \widehat{f}(k_1)2\pi\delta(k_2)$.

(2) Find the 3-dimensional Fourier transform $\widehat{f}(k_1, k_2, k_3)$ of the functions (in spherical coordinates $r,\ \theta,\ \varphi$)

$$f_1(r) = 1/r^2 \quad \text{and} \quad f_2(r) = \exp(-r)$$

Hint: Write $\exp(i\,\mathbf{k}\cdot\mathbf{x})d^3\mathbf{x} = \exp(ikr\cos\theta)\,r^2\,\sin\theta\,d\theta\,d\varphi\,dr$ (with $k = |\mathbf{k}|$, $r = |\mathbf{x}|$) and perform first the (trivial) integration in $d\varphi$, then in $d\theta$, ….

3.114

(1) Find the inverse Fourier transform in $\mathbf{R}^3$ of $g(k) = 1/k^2$ (similar calculations as in the previous problem) and deduce $\mathscr{F}(1/r)$.

(2) Using Fourier transform verify that the Green function for the Poisson equation

$$\Delta V = -4\pi\rho(\mathbf{x}), \qquad V = V(\mathbf{x}),\ \mathbf{x} \in \mathbf{R}^3$$

(where Δ is the 3-dimensional Laplacian) is just $G(r) = 1/r$. Deduce the elementary rule for finding the electric potential $V(x)$ produced by a distribution of charges $\rho = \rho(\mathbf{x})$, i.e. $V = (1/r) * \rho = \cdots$.

3.115

Verify, using Fourier transform, that the harmonic function $u(x, y) = x^2 - y^2$ satisfies the equation (see also Problem 3.113, q.(1))

$$\Delta_2 u = 0, \qquad u = u(x, y)$$

where Δ_2 is clearly the 2-dimensional Laplacian. The same question for $u(x, y) = x^3 - 3xy^2$.

3.3 Laplace Transforms

The notations for the Laplace transform of a (locally summable) function $f = f(x)$ will be

$$\mathscr{L}\big(f(x)\big) = \int_0^{+\infty} f(x)\exp(-sx)\,dx = g(s) = \widetilde{f}(s)\,, \qquad s \in \mathbf{C}$$

with $\mathrm{Re}\, s > \lambda$, *where* λ *is the* summability abscissa *of the Laplace transform. According to this definition, the functions $f(x)$ to be transformed are defined only for $x \geq 0$ (or must be put equal to zero for $x < 0$); so that, e.g., the transform $\mathscr{L}(\sin x) = 1/(1+s^2)$ should be more correctly written $\mathscr{L}\big(\theta(x)\sin x\big) = 1/(1+s^2)$. The function $\theta(x)$ is usually understood and omitted in this context; it will be explicitly introduced only when possible misunderstanding can occur, especially when comparing Fourier and Laplace transforms.*

Some other useful transforms: $\mathscr{L}\big(\exp(kx)\big) = 1/(s-k)$ with $\lambda = \mathrm{Re}\, k$; $\mathscr{L}(x^n) = n!/s^{n+1}$ with $n = 0, 1, \ldots$ and $\lambda = 0$; $\mathscr{L}\big(f(x-a)\,\theta(x-a)\big) = \exp(-sa)\mathscr{L}\big(f(x)\big)$, with $a > 0$.

An useful "summability criterion": Given $f(x)$, assume there are three constants $M > 0$, $a \in \mathbf{R}$ and $x_0 \geq 0$ such that $|f(x)| \leq M\exp(ax)$, $\forall x \geq x_0$: then $f(x)$ admits Laplace transform at least in the half plane $\mathrm{Re}\, s > a$, i.e., $\lambda \leq a$. An important property is that the Laplace transform $\widetilde{f}(s)$ is analytic in the half plane $\mathrm{Re}\, s > \lambda$.

Apart from Problems 3.129 and 3.130, all problems in this subsection can be solved using elementary Laplace transforms, as those given above, and standard properties of Laplace transform, i.e., no need of the general Laplace inversion formula (also known as Bromwich or Riemann–Fourier formula)

$$f(x) = \mathscr{L}^{-1}\big(\widetilde{f}(s)\big) = \frac{1}{2\pi i}\int_{\ell} \widetilde{f}(s)\exp(+sx)\,ds$$

where ℓ is any "vertical" line in the complex plane s from $a - i\infty$ to $a + i\infty$ with $a > \lambda$.

3.116

(1) Without trying to evaluate the Laplace transform of the following functions, specify their summability abscissas:

$$x^\alpha\,;\ x^\alpha\exp(-x^2)\,;\ x^\alpha\exp(\beta x)\,,\ \text{with}\ \alpha > -1\ \text{(why this limitation?) and}\ \beta \in \mathbf{R}$$

(2) The same question for the functions

$$\frac{1}{x+c}; \quad \frac{\exp(\gamma x)}{x+c}; \quad \frac{\sin(x^2)}{(x+c)^2}, \text{ with } c>0 \text{ (why this limitation?) and } \gamma \in \mathbf{C}$$

3.117

(1) Study the singularities in the complex plane s of the following Laplace transform

$$g(s) = \frac{\exp(-s)+s-1}{s^2(s+1)}$$

(2) Find and draw the inverse Laplace transform $f(x) = \mathscr{L}^{-1}\big(g(s)\big)$.

(3) Deduce from (1) and/or (2) the summability abscissa λ of $g(s)$.

3.118

The same questions as in the previous problem for the Laplace transform

$$g(s) = \frac{1+\exp(-\pi s)}{s^2+1}$$

3.119

(1) Show that, if the abscissa λ of a Laplace transform $\widetilde{f}(s) = \mathscr{L}\big(f(x)\big)$ satisfies $\lambda < 0$, then the Fourier transform $\widehat{f}(\omega)$ of $f(x)$ can be obtained from $\widetilde{f}(s)$ by a simple substitution. Compare e.g. the Laplace and Fourier transforms of $f(x) = \theta(x)\exp(-x)$.

(2) Find and draw the inverse Laplace transform $f(x) = \mathscr{L}^{-1}\big(\widetilde{f}(s)\big)$ of

$$\widetilde{f}(s) = \frac{1-\exp(-s)-s\,\exp(-s)}{s^2}$$

and find its abscissa λ. Compare this Laplace transform with the Fourier transform $\widehat{f}(\omega) = \mathscr{F}\big(f(x)\big)$.

(3) This and the following question deal with the "critical" case $\lambda = 0$ (see also next problem). Obtain $\mathscr{F}\big(\theta(x)\big)$ as $\lim_{\varepsilon\to 0^+} \mathscr{F}\big(\theta(x)\exp(-\varepsilon x)\big)$, $\varepsilon > 0$, and compare with $\mathscr{L}\big(\theta(x)\big)$.

(4) Find and draw the inverse Laplace transform $f(x) = \mathscr{L}^{-1}\big(\widetilde{f}(s)\big)$ of

$$\widetilde{f}(s) = \frac{1-\exp(-s)}{s^2}$$

Find then the Fourier transform $\widehat{f}(\omega) = \mathscr{F}\big(f(x)\big)$ and compare with $\widetilde{f}(s)$.

3.120

Without trying to evaluate their Laplace and Fourier transforms, show that the following functions admit $\mathscr{L}$-transform with abscissa $\lambda = 0$; do these functions also admit $\mathscr{F}$-transform (in $L^1(\mathbf{R}),\ L^2(\mathbf{R})$ or $\mathscr{S}'$)? (see also the examples in questions (3) and (4) of the problem above)

$$\theta(x)\frac{1}{1+x^2}\quad;\quad\theta(x)\frac{x}{1+x^2}\quad;\quad\theta(x)\frac{x^2}{1+x^2}\quad;\quad\theta(x)\exp(\pm\sqrt{x})$$

3.121
Using the result obtained in q. (1) of Problem 3.119:

(1) Show that if $f(x)\in L^2(I)$ where $I\subset\mathbf{R}$ is a compact interval, then its Fourier transform $\widehat{f}(\omega)$ is an analytic function for all $\omega\in\mathbf{C}$.

(2) Deduce that, in the same assumption, $\widehat{f}(\omega)$ can have at most isolated zeroes in the complex plane; in particular cannot have compact support on the real axis.

(3) Show that properties (1) and (2) are also shared, for instance, by the functions $\theta(x)\exp(-x^2)$, $\theta(x)\exp(-x^4)$ and also $\exp(-x^2),\ \exp(-x^4)$ (the argument can be easily extended to $x<0$).

3.122
Consider the Laplace transform

$$\widetilde{f}(s)\ =\ \frac{\exp(-as)-\exp(-bs)}{s^2+1}\,,\qquad a,\,b\in\mathbf{R}$$

(1) How can one choose $a,\ b\in\mathbf{R}$ in order to have abscissa $\lambda=-\infty$?

(2) Find and draw the inverse Laplace transform $f(x)=\mathscr{L}^{-1}\big(\widetilde{f}(s)\big)$ if $a=0$ and in the cases $b=\pi,\ 2\pi\ ,4\pi$.

(3) Use the above results to obtain the inverse Fourier transform of the function

$$g_{n,m}(\omega)\ =\ \frac{\exp\big(i\omega(1+2\pi m)\big)-\exp\big(i\omega(1+2\pi n)\big)}{1-\omega^2}$$

where $n,\,m\in\mathbf{Z}$ with $n>m$. Show that $g_{n,m}(\omega)$ is a C^∞ function. What is the support of this inverse Fourier transform?

3.123
The equation of an electric series circuit of a resistance R, an inductance L and a capacitor C is, with usual notations,

$$\frac{1}{C}\int_0^t I(t')\,dt'+L\frac{d}{dt}I+RI\ =\ V(t)\,,\qquad I=I(t)$$

(differently from the previous problems, here the independent variable is the time t). Transform this equation by means of Laplace transform and show that, putting $I(0)=0$, the Laplace transforms $\widetilde{V}(s)=\mathscr{L}\big(V(t)\big)$ and $\widetilde{I}(s)=\mathscr{L}\big(I(t)\big)$ are related by the rule

$$\widetilde{V}(s)=\widetilde{G}(s)\,\widetilde{I}(s)$$

where $\widetilde{G}(s)$ is the Laplace transform of a Green function $G(t)$. Find and draw $G(t)$ for different values of R, L, C (cf. Problem 3.6).

3.124
Consider the equation of a harmonic oscillator subjected to an external force $f(t)$

$$\ddot{y} + y = f(t)\,, \qquad y = y(t)$$

(as in the problem above, the independent variable is the time t), with given initial values $y(0) = a,\ \dot{y}(0) = b$.

(1) Put $f(t) \equiv 0$ and solve by means of Laplace transform the equation (the solution is trivial and well-known!)

(2) Put $a = b = 0$ and

$$f(t) = \theta(t) - \theta(t-c)\,, \qquad c > 0$$

Write the Laplace transform $\widetilde{f}(s) = \mathscr{L}\big(f(t)\big)$. Find and draw the solution $y(t)$ if $c = \pi$ and if $c = 2\pi$. For what values of c one has $y(t) = 0$ for any t larger than some $t_0 > 0$?

(3) The same questions as in (2) with

$$f(t) = \delta(t) + \delta(t-c)$$

where $\delta(t)$ is the Dirac delta.

3.125
Using $\mathscr{L}\big(xf(x)\big) = \cdots$ and $\lim_{\operatorname{Re} s \to +\infty} \mathscr{L}\big(f(x)\big) = 0$, find

$$\mathscr{L}\left(\frac{\exp(ax) - \exp(bx)}{x}\right) \quad ; \quad \mathscr{L}\left(\frac{\sin x}{x}\right)$$

3.126
(1) The Bessel function of zero order $y = J_0(x)$ satisfies the ODE

$$xy'' + y' + xy = 0$$

Applying Laplace transform to this equation, find the Laplace transform $\widetilde{J_0}(s) = \mathscr{L}\big(J_0(x)\big)$. *Hint*: use the rules $\mathscr{L}\big(xf(x)\big) = \cdots$ and $\mathscr{L}\big(f'(x)\big) = \cdots$ and obtain a first-order ODE for $\widetilde{J_0}(s)$ which can be directly solved; recall that $J_0(0) = 1$ and the "initial value theorem":

$$f(0^+) = \lim_{x \to 0^+} f(x) = \lim_{\operatorname{Re} s \to +\infty} s\,\mathscr{L}\big(f(x)\big)$$

(2) As an application, find the convolution product $J_0(x) * J_0(x)$.

3.127

In the following questions, use a translation rule for Laplace transforms and the known formula

$$\sum_{n=0}^{\infty} \exp(-nas) = \frac{1}{1-\exp(-as)}, \qquad a>0,\ \mathrm{Re}\, s>0$$

(*a*) Let $f_0(x)$ be a given function in $0<x<T$ and $=0$ elsewhere, and let $f(x)$ be the function obtained repeating periodically for $x>T$ the function $f_0(x)$ with period T. Let $F_0(s)=\mathscr{L}\big(f_0(x)\big)$. Obtain $F(s)=\mathscr{L}\big(f(x)\big)$. Find then $\mathscr{L}\big(f(x)\big)$ where $f(x)$ is the square wave function:

$$f(x) = \begin{cases} 1 & \text{for } 0<x<1,\ 2<x<3,\ \ldots,\ 2n<x<2n+1,\ldots \\ -1 & \text{for } 1<x<2,\ 3<x<4,\ldots \end{cases}$$

(*b*) Find $\mathscr{L}^{-1}\big(g(s)\big)$ where (see Problem 3.118)

$$g(s) = \frac{1}{s^2+1}\frac{1+\exp(-\pi s)}{1-\exp(-\pi s)} = \frac{1}{s^2+1}\mathrm{ctanh}\big(\pi s/(2)\big)$$

(*c*) Find $\mathscr{L}^{-1}\Big(\dfrac{1}{s(1-\exp(-s))}\Big)$

3.128

Consider the d'Alembert equation

$$\frac{\partial^2 u}{\partial x^2} = \frac{\partial^2 u}{\partial t^2}, \qquad u=u(x,t) \quad .$$

with given initial conditions

$$u(x,0) = f(x)\,, \quad u_t(x,0) = g(x)$$

Introduce and use the Laplace transform $\widetilde{u}(x,s)$ with respect to t, i.e.

$$\widetilde{u}(x,s) = \int_0^{+\infty} u(x,t)\exp(-st)\,dt$$

(1) Show that the d'Alembert equation is transformed into an ODE for $\widetilde{u}(x,s)$.

(2) In the case of vanishing initial conditions $f(x)=g(x)=0$, solve the ODE obtained in (1) to obtain the Laplace transform $\widetilde{u}(x,s)$ of the most general solution of the equation.

(3) In the same conditions as in (2), assume that the equation describes an elastic semi-infinite string (in $x\geq 0$), and that extremum at $x=0$ of the string is subjected to

a given transversal displacement $u(0, t) = \varphi(t)$. Find the solution $u(x, t)$ imposing the condition that $u(x, t)$ is bounded for all $x \geq 0$.

3.129

Using the general Laplace inversion formula (see the Introduction):

(1) Show that the result $f(x)$ is independent of the choice of the abscissa a of the line ℓ of integration (provided that $a > \lambda$) and that $f(x) = 0$ if $x < 0$, as expected.

(2) Verify, by means of integration in the complex plane s, the well-known formula

$$\mathscr{L}^{-1}\left(\frac{1}{s^2+1}\right) = \theta(x) \sin x$$

3.130

Use the general Laplace inversion formula (see the Introduction) to calculate

$$f(x) = \mathscr{L}^{-1}\left(\frac{1}{\sqrt{s}}\right)$$

Hint: An integration in the complex plane s is requested in the presence of a cut to be placed along the negative part of the real axis $s_1 = \mathrm{Re}\, s$.

3.131

(1) Using Laplace transform, show that the set $\{u_n(x) = x^n \exp(-x),\ n = 0, 1, 2, \ldots\}$ is a complete set in the Hilbert space $L^2(0, +\infty)$. *Hint*: show first that the Laplace transform of a function $f(x) \in L^2(0, +\infty)$ has abscissa $\lambda \leq 0$. Verify then that the completeness condition $(u_n, f) = 0,\ \forall n = 0, 1, 2, \ldots$ becomes a condition on the derivatives of the Laplace transform $\widetilde{f}(s)$ evaluated at $s = \cdots$.

(2) Proceeding as before, show that the set $\{u_n(x) = x^n \exp(-x) \exp(-x/n),\ n = 1, 2, \ldots\}$ is a complete set in the Hilbert space $L^2(0, +\infty)$. *Hint*: use a property of the set of zeroes of an analytic function.

Chapter 4
Groups, Lie Algebras, Symmetries in Physics

4.1 Basic Properties of Groups and of Group Representations

Most of the physical applications of group theory are based on the notion of group representation. After some basic preliminaries, the main definitions and properties are shortly summarized below.

A subgro up H of a group G is said to be invariant (or normal) if $g\,h\,g^{-1} \in H$, $\forall h \in H$, $\forall g \in G$. A group is said to be simple if it does not contain invariant subgroups (apart from, obviously, the identity and the group itself), and it is semisimple *if it does not contain Abelian (i.e., commutative) invariant subgroups. These definitions can be easily extended to Lie algebras, with the notions of invariant subalgebras and Abelian invariant subalgebras.*

Given two groups G, G', a homomorphism Φ is a map $\Phi : G \to G'$ such that (using a multiplicative notation for both groups) $\Phi(g_1 \cdot g_2) = \Phi(g_1) \cdot \Phi(g_2)$, $\forall g_1, g_2 \in G$. The kernel of any homomorphism is an invariant subgroup of G.

Given a group G and a vector space V (either on **R** *or* **C***, either a finite-dimensional or a Hilbert space), a (linear)* representation *$\mathscr{R}$ of G is a homomorphism $\mathscr{R} : G \to GL(V)$ of G in the group of the linear transformations of V in itself. If the dimension n of V is finite, chosen a (orthonormal) basis in V (now, $V \equiv \mathbf{R}^n$ or $V \equiv \mathbf{C}^n$), this defines a $n \times n$ matrix representation of the group. Two representations $\mathscr{R}$ and $\mathscr{R}'$ are equivalent if there is a non-singular matrix S such that $\mathscr{R}'(g) = S\,\mathscr{R}(g)\,S^{-1}$, $\forall g \in G$,* i.e., *if $\mathscr{R}$ and $\mathscr{R}'$ simply differ in a change of basis. A representation is (partially)* reducible *if there is a proper invariant subspace $V_1 \subset V$, i.e., if $\mathscr{R} : V_1 \to V_1$; the representation $\mathscr{R}$* is completely reducible *if there is also an invariant subspace V_2, complemen-*

G. Cicogna, *Exercises and Problems in Mathematical Methods of Physics*, Undergraduate Lecture Notes in Physics,
https://doi.org/10.1007/978-3-030-59472-5_4

tary to V_1, *i.e., if* $\mathscr{R} : V_2 \to V_2$. *A representation* $\mathscr{R} : V \to V$ is irreducible *if no such subspaces exist. A representation is* faithful *if is injective.*

Two elements $g_1, g_2 \in G$ *are said to be* conjugate *if there is* $g' \in G$ *such that* $g_2 = g' g_1 g'^{-1}$; *then, the group can be partitioned into conjugacy classes (in each class the elements are conjugated to one another). If the order N of the group (i.e., the number of the elements contained in the group) is finite, then, according to Peter-Weyl and Burnside theorems, the number of the* inequivalent irreducible representations *of the group is equal to the number s of conjugacy classes, and their dimensions* $d_1, \ldots, d_s$ *satisfy the Burnside theorem*

$$d_1^2 + d_2^2 + \ldots + d_s^2 = N$$

Given any matrix representation $\mathscr{R}$ *of a group G, the character* $\chi_{\mathscr{R}}$ *of* $\mathscr{R}$ *is defined as the set of the traces of the matrices* $\mathscr{R}(g)$:

$$\chi_{\mathscr{R}}(g) = \mathrm{Tr}\, \mathscr{R}(g)$$

If the order N of the group G is finite, the characters are N-dimensional vectors in $\mathbf{R}^N$ (or in $\mathbf{C}^N$), *and the characters of the inequivalent irreducible* representations $\mathscr{R}_a$ *of G are orthogonal, more precisely one has*

$$\left(\chi_{\mathscr{R}_a}, \chi_{\mathscr{R}_b}\right) = N\,\delta_{ab}\,, \quad a, b = 1, \ldots, s$$

where $(\,,\,)$ *is the scalar product in* $\mathbf{R}^N$ (or in $\mathbf{C}^N$).

Let $\mathscr{R}$ *be a reducible representation of a group and* $\mathscr{R}_a$, $a = 1, \ldots, s$, *its* inequivalent irreducible representations; *the decomposition of* $\mathscr{R}$ *can be written*

$$\mathscr{R} = n_1 \mathscr{R}_1 \oplus \ldots \oplus n_s \mathscr{R}_s$$

where n_a *is the number of times the irreducible representation* $\mathscr{R}_a$ *occurs in* $\mathscr{R}$. *Using the orthogonality of the characters, the numbers* n_a *are given by*

$$n_a = N^{-1}\left(\chi_{\mathscr{R}}(g), \chi_{\mathscr{R}_a}(g)\right)$$

4.1

Consider a group G of finite order N:

(1) Show that the order of all subgroups of G is a divisor of N. *Hint:* let H be a subgroup; consider the cosets[1] gH and verify that they are disjoint and one-to-one with H, ….

(2) Deduce that if N is a prime number then G has no *proper* subgroups (i.e., apart from the identity e and G itself), and therefore, in particular, is simple.

[1] For any fixed $g \in G$, the coset gH is the set $\{gh,\ h \in H\}$.

(3) Show that if N is prime, then G is cyclic and Abelian. *Hint:* for any $g \in G$, the set $\{e, g, g^2, \ldots\}$ is a subgroup of G, ...

4.2
Show that if $\mathscr{R}$ is an *unitary* representation of a group G acting on a basis space V (V may be either finite-dimensional or a Hilbert space) and $\mathscr{R}$ admits an invariant subspace $V_1 \subset V$, then also the orthogonal complementary subspace V_2 (i.e. $V = V_1 \oplus V_2$) is invariant under $\mathscr{R}$. Therefore, $\mathscr{R}$ is completely reducible: $\mathscr{R} = \mathscr{R}_1 \oplus \mathscr{R}_2$ with $\mathscr{R}_i : V_i \to V_i$.

4.3
Let $\mathscr{R}$ be any representation of a group G on a basis space V (V may be either finite dimensional or a Hilbert space) and let T be any operator defined in V and commuting with $\mathscr{R}$, i.e.,

$$T\,\mathscr{R}(g) = \mathscr{R}(g)\,T\,, \qquad \forall g \in G$$

(1)*(i)* Let λ be an eigenvalue of T and let V_λ be the subspace of the eigenvectors of T with eigenvalue λ. Show that V_λ is invariant under $\mathscr{R}$, i.e. that if $v_\lambda \in V_\lambda$ then also $v' \equiv \mathscr{R}(g)\,v_\lambda \in V_\lambda$, $\forall g \in G$, or $Tv' = \lambda v'$.

(ii) If, in particular, $\mathscr{R}$ is irreducible, deduce the Schur lemma: $T = \lambda\, I$ (in its simplest version).

(2) Show the following generalization: Let $\mathscr{R}$ be a completely reducible representation $\mathscr{R} = \mathscr{R}_1 \oplus \mathscr{R}_2 \oplus \ldots$ where $\mathscr{R}_a$ are inequivalent irreducible representations acting on the subspaces V_a with $V = V_1 \oplus V_2 \oplus \ldots$. If $T\mathscr{R}(g) = \mathscr{R}(g)T$, then $T = \lambda_1 P_1 + \lambda_2 P_2 + \ldots$, where P_a projects on the subspace V_a. *Hint*: Use this property: if $\mathscr{R}_1$ and $\mathscr{R}_2$ are two inequivalent irreducible representations and A is a matrix satisfying $\mathscr{R}_1(g)A = A\,\mathscr{R}_2(g)$, $\forall g \in G$, then $A = 0$.

(3) Let now T be an operator which "respects the symmetry" of a system: in other words, T remains unchanged under the transformations of the group G describing the symmetry of the system, i.e., $T = \mathscr{R}(g)\,T\,\mathscr{R}(g^{-1})$, $\forall g \in G$, or $T\mathscr{R}(g) = \mathscr{R}(g)T$ (see Problems 4.5–4.8, 4.36, 4.38–4.39). Conclude: if a system admits a symmetry group G, what degeneracies can be expected, in general, for the eigenvalues of T?

4.4
(1) Show that if a group G is Abelian then all its irreducible representations are 1-dimensional (it is understood that we are operating in complex spaces: see Problem 4.11,q.(1)).

(2) Show that if a group G is simple then all its representations are faithful (apart from, obviously, the trivial representation $\mathscr{R} : g \to 1$, $\forall g \in G$).

4.5
(1) Find the group describing the symmetry of the equilateral triangle, find its conjugacy classes and all its inequivalent irreducible representations. What degeneracy can be expected for the eigenvalues of any operator respecting the symmetry (see

Problem 4.3, q.(3) and next problem) of a system which exhibits this symmetry (e.g.., the molecule of ammonia NH_3)?

(2) Find the characters of the inequivalent irreducible representations of the symmetry group of the triangle and verify their orthogonality property.

4.6

(1) Consider a system consisting of three particles with the same mass m, placed at the vertices of an equilateral triangle with side ℓ, and subjected to elastic forces produced by three equal springs placed along the sides of the triangle. The length at rest of the springs is ℓ. To describe the small displacements (in the plane) of the particles from their equilibrium positions one clearly needs 6-dimensional vectors $\mathbf{x} \in \mathbf{R}^6$, where, e.g.., the vectors in the subspace generated by the first two basis vectors e_1, e_2 describe the displacements of the first mass, the vectors in the subspace e_3, e_4 those of the second mass, etc. The action of the symmetry group of the triangle on this space produces then a 6-dimensional representation $\mathscr{R}$. Show that this 6-dimensional representation has the following character

$$\chi_{\mathscr{R}} = (6; 0, 0; 0, 0, 0)$$

To obtain this character, notice that only diagonal elements must be evaluated, then for instance the diagonal elements of the rotations are all zero, because any rotation changes e_1, e_2 into e_3, e_4 or e_5, e_6, etc. Using the characters of the inequivalent irreducible representations of the group of the triangle (previous problem), decompose then this representation in a direct sum of irreducible representations. These describe different displacements of the particles. Two of these displacements are expected to be trivial, corresponding to rigid displacements: translations (2 dimensions) and rotations (1 dimension), and the other two, having dimension 1 and 2, correspond to oscillations of the system. Let T be a 9×9 matrix describing the mutual elastic interactions of the particles; although the explicit form of T is not explicitly given, it must respect the symmetry, according to Problem 4.3, q.(3), i.e., $T\mathscr{R} = \mathscr{R}T$. The normal modes of the system $\mathbf{x} = \mathbf{a}\cos(\omega t + \varphi)$, $\mathbf{a} \in \mathbf{R}^6$, satisfy an equation as $T\,\mathbf{x} = \ddot{\mathbf{x}}$, or $T\,\mathbf{a} = -\omega^2\mathbf{a}$. What about the degeneracies of the frequencies of these oscillating modes?

(2) What effect can be expected if a "small" perturbation breaking the equilateral triangle symmetry is introduced (e.g.., changing the mass of one particle)?

4.7

Consider the group of symmetry of the square: determine its order N, its conjugacy classes and the dimensions of its inequivalent irreducible representations. Compare with the result obtained in Problem 1.101 about the degeneracy of the eigenvalues of the Laplace operator $T = \partial^2/\partial x^2 + \partial^2/\partial y^2$ for the square with vanishing boundary conditions. Show that "accidental degeneracies" are present in this case.

4.8

The symmetry group $\mathscr{O}$ of the cube contains a subgroup (denoted here by $\mathscr{O}_1$) of 24 transformations not involving reflections, and other 24 transformations including reflections. Show that the group can be written as a direct product $\mathscr{O} = \mathscr{O}_1 \times Z_2$, where Z_2 is (isomorphic to) $\{1, -1\}$. Show that there are 5 conjugacy classes in $\mathscr{O}_1$ and, using also Burnside theorem, find the dimensions of the inequivalent irreducible representations of $\mathscr{O}_1$ (and of those of $\mathscr{O}$), and the degeneracy which can be expected for the eigenvalues of any operator respecting the symmetry of the cube.

4.9

(1) Find all the inequivalent irreducible representations of the additive group Z_7 of the integers *mod 7*. *Hint*: see Problems 4.1, q. (3), 4.4, q. (1) and recall Burnside theorem.

(2) The same for the group Z_6. Are there not-faithful representations? Compare with the case of Z_7 and with the symmetry group of the equilateral triangle (Problem 4.5).

4.10

(1) Let H be an invariant subgroup of a group G: show that the set of the cosets gH can be naturally equipped with a structure of group: the quotient group $Q = G/H$ (notice that H may be the kernel K of any homomorphism $\Phi : G \to G'$, or of any representation $\mathscr{R}$ of G). It is not true, in general, that Q is an invariant subgroup of G; but if the elements contained in Q commute with those in H, then G can be written as a *direct product* $G = Q \times H$: see next questions.

(2) Let Φ be the homomorphism $\Phi : GL_n(\mathbf{C}) \to \mathbf{C}$ of the general linear complex group $GL_n(\mathbf{C})$ of the invertible $n \times n$ complex matrices M into the multiplicative group of nonzero complex numbers $\mathbf{C}$, defined by

$$\Phi(M) = \det M$$

Specify the kernel $K = \mathtt{Ker}\,\Phi$ and the quotient group $Q = GL_n(\mathbf{C})/K$. Is it true that $GL_n(\mathbf{C})$ is the direct product $GL_n(\mathbf{C}) = K \times Q$?

(3) The same questions for the group U_n of the unitary matrices.

(4) The same questions for the group $GL_n(\mathbf{R})$ of the invertible real matrices, and for the group O_n of the orthogonal matrices.

4.11

(1) The "basic" representation of the rotation group SO_2 in $\mathbf{R}^2$ is, as well-known,

$$R(\varphi) = \begin{pmatrix} \cos\varphi & -\sin\varphi \\ \sin\varphi & \cos\varphi \end{pmatrix}$$

Clearly, no vector v in the *real* space $\mathbf{R}^2$ satisfies of $R(\varphi)v = \lambda v$ (apart from the case $\varphi = \pi$), but there *must* exist eigenvectors in $\mathbf{C}^2$ (SO_2 is Abelian, then its irreducible representations are 1-dimensional): find the eigenvectors of $R(\varphi)$ and then decompose the representation. What can be a physical meaning of these eigenvectors in $\mathbf{C}^2$? (for a possible interpretation, see q. (3)).

(2) What are the other inequivalent irreducible representations of SO_2? What of these are faithful? For comparison: what are the representations of O_2?

(3) The electric field **E** of a planar electromagnetic wave propagating along the z-axis can be written, with usual and clear notations,

$$\mathbf{E}(x, y, z, t) = \begin{pmatrix} E_x \\ E_y \end{pmatrix} =$$

$$= \begin{pmatrix} E_1 \cos(kz - \omega t + \varphi_1) \\ E_2 \cos(kz - \omega t + \varphi_2) \end{pmatrix} = \mathrm{Re}\left\{\exp\left(i(kz - \omega t)\right) \begin{pmatrix} E_1 \exp(i\varphi_1) \\ E_2 \exp(i\varphi_2) \end{pmatrix}\right\}$$

Then the vector $\mathbf{p} \in \mathbf{C}^2$ defined by

$$\mathbf{p} = \begin{pmatrix} E_1 \exp(i\varphi_1) \\ E_2 \exp(i\varphi_2) \end{pmatrix}$$

describes the *state of polarization* of the e.m. wave. E.g., $\mathbf{p} = (1, 0)$ describes the linear polarization along the x-axis, etc. What types of polarization are described by $\mathbf{p} = (1, \pm i)$?

4.12

Verify that the Lorentz boost with velocity v involving one spatial variable x and the time t can be written in the form

$$L(\alpha) = \begin{pmatrix} \cosh\alpha & -\sinh\alpha \\ -\sinh\alpha & \cosh\alpha \end{pmatrix}$$

where $\alpha = \arctan(v/c)$. This (not unitary!) representation of the "pure" (i.e., without space and time inversions) 1-dimensional Lorentz group is reducible (the group is indeed Abelian): perform this reduction; what is the physical meaning of the eigenvectors of $L(\alpha)$?

4.2 Lie Groups and Lie Algebras

Thanks to Ado theorem, which essentially states that all finite dimensional Lie algebras admit a faithful representation by means of matrices, we will adopt the "concrete" use of matrices to describe Lie groups and Lie algebras: see Problems 4.13–4.17. *Starting with Problem* 4.18, *we will also use the description of Lie algebras by means of differential operators.*

4.13

(1) Show that the matrices $M \in SU_n$ can be put in the form

$$M = \exp A$$

where the matrices A are anti-Hermitian (i.e., $A^+ = -A$) and traceless. Show that the space $\mathscr{A}$ of these matrices is a vector space and that for any $A_1,\ A_2 \in \mathscr{A}$ then also the commutator $[A_1, A_2] \in \mathscr{A}$. Find the dimension of the space $\mathscr{A}$, the Lie algebra of SU_n, as vector space on the reals (which is, by definition, the dimension of SU_n); find then the dimension of U_n.

(2) Show that the matrices $M \in SO_n$ can be put in the form $M = \exp A$, where the matrices A are real antisymmetric. Notice however that, in this case, imposing *only* the orthogonality of M, one gets automatically *also* $\mathrm{Tr}\,A = 0$ and then $\det M = 1$: explain why this procedure cannot be extended to the whole group O_n, i.e. to the matrices with $\det = -1$. Find the dimension of the groups SO_n. In the case of SO_3, find a basis for the vector space $\mathscr{A}$ of the 3×3 real antisymmetric matrices.

4.14

Consider a group G of matrices M and the neighborhood of the identity where one can write (as done, e.g.., in the previous problem) $M = \exp A$, where the matrices A describe a vector space $\mathscr{A}$, the Lie algebra of G, of dimension r (over the reals).

(1) Choose a matrix $A_1 \in \mathscr{A}$ and consider $M_1 = \exp(a_1 A_1)$ where $a_1 \in \mathbf{R}$. Show that M_1 is an Abelian one-parameter subgroup of G. Let M_1 and $M_2 = \exp(a_2 A_2)$ two of these subgroups: in general, do these subgroups commute?

(2) Let $A_1, \ldots, A_r$ be a basis for the vector space $\mathscr{A}$: any $A \in \mathscr{A}$ can then be written as $A = \sum_{i=1}^{r} a_i A_i$ with $\mathbf{a} \in \mathbf{R}^r$; show that

$$A_i = \left.\frac{\partial M}{\partial a_i}\right|_{\mathbf{a}=0}$$

4.15

(1) Find the expression of the generator A of the Lie algebra of the rotation group SO_2 in its "basic" representation given in Problem 4.11, q.(1).

(2) The same for the pure Lorentz group given in Problem 4.12.

(3) Evaluate

$$\exp(aA) = \sum_{n=0}^{\infty} \frac{(aA)^n}{n!}$$

in the cases

$$A = \begin{pmatrix} 0 & -1 \\ 1 & 0 \end{pmatrix} \; ; \; A = \begin{pmatrix} i & 0 \\ 0 & -i \end{pmatrix} \; ; \; A = \begin{pmatrix} 0 & -1 \\ -1 & 0 \end{pmatrix} \; ; \; A = \begin{pmatrix} 1 & 0 \\ 0 & -1 \end{pmatrix}$$

Hint: $A^2 = \ldots$. Notice that this is a partial converse of questions (1) and (2).

4.16

Describe (or give a geometrical or physical interpretation of) the groups generated by the following (1-dimensional) Lie algebras:

$$A = 1; \quad A = i; \quad A = \begin{pmatrix} 1 & 0 \\ 0 & 2 \end{pmatrix}; \quad A = \begin{pmatrix} i & 0 \\ 0 & 2i \end{pmatrix}; \quad A = \begin{pmatrix} i & 0 \\ 0 & i\sqrt{2} \end{pmatrix};$$

$$A = \begin{pmatrix} 1 & -1 \\ 1 & 1 \end{pmatrix}; \quad A = \begin{pmatrix} 0 & -1 & 0 \\ 1 & 0 & 0 \\ 0 & 0 & 1 \end{pmatrix}; \quad A = \text{any } n \times n \text{ matrix}$$

4.17

The elements of the Euclidean group E_3 in $\mathbf{R}^3$ (or in general E_n in $\mathbf{R}^n$) are the transformations $\mathbf{x} \to \mathbf{x}' = \mathscr{O}\mathbf{x} + \mathbf{b}$, where $\mathscr{O} \in O_3$ and $\mathbf{b} \in \mathbf{R}^3$. Write the composition rule in E_3 and show that the mapping

$$(\mathscr{O}, \mathbf{b}) \to \begin{pmatrix} & & & b_1 \\ & \mathscr{O} & & b_2 \\ & & & b_3 \\ 0 & 0 & 0 & 1 \end{pmatrix}$$

is a faithful (partially reducible) representation of E_3 in the group of 4×4 matrices. Find the expression of the Lie generators $A_1,\ A_2,\ A_3$ and $B_1,\ B_2,\ B_3$ of E_3 in this representation, where A_i generate the rotations and B_i the translations (start with the rotations around the z-axis).

4.18

(1) *(a)* Let T_1 be the group of translations along $\mathbf{R}$ and let the parameter $a \in \mathbf{R}$ denote the translation (clearly T_1 is isomorphic to $\mathbf{R}$, the additive group of reals). Consider the representation of T_1 acting on the functions $f(x) \in L^2(\mathbf{R})$ according to

$$a \to U_a \quad \text{where} \quad (U_a f)(x) = f(x - a)$$

Show that, as suggested by the notation, this representation is unitary. Assume the parameter a "infinitesimal" and $f(x)$ regular enough (expandable); expand then

$$f(x - a) = f(x) - \ldots$$

and obtain the differential expression A of the Lie generator of the translation group. Notice that the generator is anti-Hermitian,[2] in agreement with Stone theorem. To be precise, Stone theorem requires also the *strong continuity* of the representation,

---

[2]It is customary in physics to introduce a factor i in the definition of these operators, in order to have *Hermitian* operators. E.g., $A = -i\,d/dx$, which is proportional to the momentum operator $P = -i\,\hbar d/dx$ in quantum mechanics, as well-known.

which means in this case that, given two translations U_a, U_b, if $b \to a$ then $\|U_a f - U_b f\| \to 0$. This is indeed is true, as shown in Problem 3.65, q.(4).

(b) Find the generators of the other three representations (α), (γ), (δ) of the translation group proposed in the next problem. What of these are anti-Hermitian?

(c) Extend to the group of translations in $\mathbf{R}^3$, i.e. $\mathbf{x} \to \mathbf{x}' = \mathbf{x} + \mathbf{a}$, $\mathbf{a} \in \mathbf{R}^3$.

(2) Consider the representation $\mathscr{R} = \mathscr{R}_\varphi$ of the rotation group SO_2 on the plane $\mathbf{x} \equiv (x, y)$ acting on the space of functions $f(\mathbf{x}) \in L^2(\mathbf{R}^2)$ according to the usual rule

$$f(\mathbf{x}) \to \big(\mathscr{R}_\varphi f\big)(\mathbf{x}) = f\big(R_\varphi^{-1}\mathbf{x}\big) = f(\mathbf{x}')$$

where R_φ is given in Problem 4.11, q. (1). Consider an "infinitesimal" rotation $R_\varphi^{-1}\mathbf{x} = \mathbf{x}' \equiv (x + \varphi y + \ldots,\ y - \varphi x + \ldots)$ and assume $f(\mathbf{x})$ regular; expand then

$$f(\mathbf{x}') \equiv f(x', y') = f(x, y) + \ldots$$

and obtain the differential expression A of the Lie generator of the rotation group.

(3) Repeat the calculations considering a function $f(x, t)$ and an "infinitesimal" Lorentz boost $L(\alpha)$ (see Problem 4.12) in the plane x, t.

4.19

The representations of the non-compact group T_1 of the 1-dimensional translations along $\mathbf{R}$ cannot be simultaneously unitary, irreducible and faithful. Specify what among the above properties are (or are not) satisfied by each one of the four representations of T_1 listed below. Denote by a the translation specified by the parameter $a \in \mathbf{R}$ (T_1 is clearly isomorphic to the additive group of reals).

(α) Consider the representation by means of 2×2 matrices according to

$$a \to \begin{pmatrix} 1 & a \\ 0 & 1 \end{pmatrix}$$

(verify that this is indeed a representation of T_1: see Problem 4.17)

(β) Consider the representation acting on the functions $f(x) \in L^2(\mathbf{R})$ according to

$$a \to U_a \quad \text{where} \quad (U_a f)(x) = f(x - a)$$

(given in the previous problem, q. (1)*(a)*). In this case, introduce Fourier transform $\mathscr{F}\big(f(x)\big) = \widehat{f}(k)$, then $\mathscr{F}\big(f(x - a)\big) = \ldots$ and show that *any* subspace of functions $f(x)$ having Fourier transform $\widehat{f}(k)$ with support $J \subset \mathbf{R}$, properly contained in $\mathbf{R}$, is a (∞ - dimensional) invariant subspace for this representation. This implies that, in order to have 1-dimensional representations, one must enlarge the choice of the basis space and consider "functions" having Fourier transform $\widehat{f}(k)$ with support in a single point, i.e. $\widehat{f}(k) = \delta(k - \lambda)$, where λ is any real number. This leads to the next representation (γ) proposed.

(γ) Consider the representation

$$a \to \exp(i\lambda a) \qquad \text{(a representation for each fixed } \lambda \in \mathbf{R}\text{)}$$

(δ) Consider the representation

$$a \to \exp(\mu a) \quad \text{(a representation for each fixed, not purely imaginary, } \mu \in \mathbf{C}\text{)}$$

4.20

(1) Extend the representations considered in the previous problem to the group of translations including reflections (i.e., $Sx = -x$; is this group Abelian?).

(2) The same for the group of translations in $\mathbf{R}^3$ without reflections and for the group which includes reflections.

4.21

(1) Let $x \in \mathbf{R}$ and consider a dilation $x \to x' = (\exp a)x$, with $a \in \mathbf{R}$; introduce the representation of the group of dilations acting on functions $f(x) \in L^2(\mathbf{R})$ according to the usual rule (cf. Problem 4.18)

$$a \to T_a \ \text{ where } \big(T_a f\big)(x) = f\big(\exp(-a)x\big) = f(x - ax + \ldots) = f(x) - \ldots$$

Assuming $f(x)$ regular (expandable), find the Lie generator D of the dilation group in this representation T_a. Show that this representation is not unitary and that D is not anti-Hermitian (nor Hermitian).

(2) Consider the group of transformations $a \to S_a$, depending on a parameter $a \in \mathbf{R}$, defined by

$$\big(S_a f\big)(x) = \exp(-a/2) f\big(\exp(-a)x\big) = (1 - a/2 + \ldots) f(x - ax + \ldots) = f(x) - \ldots$$

and find the generator $\widetilde{D}$ of this transformation. Show that S_a is unitary and that $\widetilde{D}$ is anti-Hermitian, in agreement with Stone theorem. (This is a reformulation of the Problem 1.100).

4.22

(1) Show that all 2-dimensional algebras can be put in one of the two forms $[A_1, A_2] = 0$ or $[A_1, A_2] = A_2$.

(2) Study the 2-dimensional algebra generated by D, where D is the generator of dilations considered in the above problem, q.(1) and the generator $P = -d/dx$ of the translations along x, and verify that this algebra admits a faithful representation by means of the two matrices (cfr. also Problem 4.17).

$$D' = \begin{pmatrix} 1 & 0 \\ 0 & 0 \end{pmatrix} \quad , \quad P' = \begin{pmatrix} 0 & 1 \\ 0 & 0 \end{pmatrix}$$

4.23

Let $D_1,\ D_2$ be the Lie generators of the dilations along x and y in the $\mathbf{R}^2$ plane, and let A be the generator of the rotations in this plane. Using either the 2×2 matrix representation or the differential representation (see Problems 4.15, 4.18, 4.21, q.(1), and 4.22) of these generators, show that they generate a 4-dimensional algebra: what is the fourth generator?

4.24

(1) Using, e.g.., the differential representation (see Problem 4.18), construct the Lie algebra (i.e., obtain the commutation rules) of the following 3-dimensional groups:

(i) the Euclidean group E_2; *(ii)* the Poincaré group consisting of the Lorentz boosts involving x and t and the two translations along x and t.

Verify that these algebras admit a 2-dimensional Abelian invariant subalgebras.

(2) The same question for the groups:

(i) SO_3; *(ii)* the Lorentz boosts involving t with two real variables $x,\ y$ and of the rotations on the plane x, y.

(3) The same question for the 6-dimensional "pure" Lorentz group[3] $\mathscr{L}$ of the Lorentz transformations involving t with the space variables in $\mathbf{R}^3$ and of the rotations in $\mathbf{R}^3$ (the latter is clearly the subgroup $SO_3 \subset \mathscr{L}$)

(4) The same question for the 10-dimensional Poincaré group, consisting of the Lorentz group and the 4 translations along $\mathbf{x}$ and t.

4.25

(1) Show that the operators

$$D_0 = \frac{d}{dx} \quad , \quad D_1 = x\frac{d}{dx} \quad , \quad D_2 = x^2\frac{d}{dx}$$

generate a 3-dimensional Lie algebra. Verify also that no (finite-dimensional) algebra can be generated by two or more operators of the form $x^n(d/dx)$ if, together with d/dx, at least one of the (integer) exponents n is > 2.

(2) Put

$$A_1 = \frac{1}{\sqrt{2}}\big(-D_0 + D_2/2\big)\ ,\ A_2 = D_1\ ,\ A_3\frac{1}{\sqrt{2}}\big(D_0 + D_2/2\big)$$

find the commutation rules $[A_i,\ A_j]$ and compare with the commutation rules of the 3-dimensional algebras mentioned in the previous problem.

4.26

Consider the 3-parameters group (called Heisenberg group) of the 3×3 matrices defined by

---

[3] $\mathscr{L}$ is the subgroup "connected with the identity" of the full Lorentz group, usually denoted by $O(3, 1)$, which includes also space inversions and time reversal. The same remark holds for the group considered in q.(2) *(ii)*, which is a subgroup of $O(2, 1)$.

$$M(\mathbf{a}) = \begin{pmatrix} 1 & a_1 & a_3 \\ 0 & 1 & a_2 \\ 0 & 0 & 1 \end{pmatrix}$$

where $\mathbf{a} \equiv (a_1, a_2, a_3) \in \mathbf{R}^3$.

(1) Find the generators $A_1,\ A_2,\ A_3$ of this group and verify that the commutation rules are

$$[A_1,\ A_2] = A_3 \quad , \quad [A_1,\ A_3] = [A_2,\ A_3] = 0$$

(which are the same – apart from the factor $i\hbar$ – as those of the quantum mechanical operators $p,\ q$ and the identity I), and find the 1-parameter subgroups generated by $A_1,\ A_2$ and A_3.

(2) Show that the map

$$M(\mathbf{a}) \to U_{\mathbf{a}} \quad \text{where} \quad \big(U_{\mathbf{a}} f\big)(x) = \exp(-ia_3)\exp(ixa_2) f(x - a_1)$$

where $f(x) \in L^2(\mathbf{R})$, is a *unitary* representation of the Heisenberg group. Find the generators of the group in this representation and verify that (expectedly!) they satisfy the same commutation rules as seen in (1).

(3) Verify that the representation given in (2) is irreducible. *Hint*: introduce Fourier transform $\widehat{f}(k) = \mathscr{F}\big(f(x)\big)$ as in Problem 4.19, case (β); notice however the strong difference between these two cases.

4.3 The Groups SO_3, SU_2, SU_3

4.27

(1) Let $\mathbf{x} \equiv (x, y, z) \in \mathbf{R}^3$ and consider the vector space generated by the quadratic monomials $x^2, y^2, z^2, xy, xz, yz$. This space is clearly invariant under the rotations $\mathbf{x} \to \mathbf{x}' = R\mathbf{x}$ with $R \in SO_3$. Decompose this space in a direct sum of subspaces which are the basis of irreducible representations of SO_3. What is the relationship between these subspaces and the spherical harmonics $Y_{\ell,m}(\theta, \varphi)$?

(2) The same for the space generated by the cubic monomials $x^3, x^2y, \ldots$.

4.28

(1) Given two Lie algebras $\mathscr{A}$ and $\mathscr{A}'$ with the same dimension r, and their generators $A_1, \ldots, A_r$ and resp. $A'_1, \ldots, A'_r$, assume that these generators satisfy the same commutation rules apart from a multiplicative nonzero constant in their structure constants, i.e.,

$$[A_i\,,\ A_j\,] = c_{ijk} A_k\,,\ \ [A'_i\,,\ A'_j\,] = \lambda\, c_{ijk} A'_k\,, \qquad \lambda \neq 0$$

Show that the generators A'_i can be redefined, simply introducing generators A''_i proportional to A'_i, in such a way that the structure constants of A_i and A''_i are the same.

(2) A commonly used choice for the generators of the algebras of SO_3 and SU_2 are

$$A_1 = \begin{pmatrix} 0 & 0 & 0 \\ 0 & 0 & -1 \\ 0 & 1 & 0 \end{pmatrix} \ ; \ A_2 = \begin{pmatrix} 0 & 0 & 1 \\ 0 & 0 & 0 \\ -1 & 0 & 0 \end{pmatrix} \ ; \ A_3 = \begin{pmatrix} 0 & -1 & 0 \\ 1 & 0 & 0 \\ 0 & 0 & 0 \end{pmatrix}$$

and resp. (the well-known Pauli matrices σ_j are given by $\sigma_j = iA_j$)

$$A'_1 = \begin{pmatrix} 0 & -i \\ -i & 0 \end{pmatrix} \ ; \ A'_2 = \begin{pmatrix} 0 & -1 \\ 1 & 0 \end{pmatrix} \ ; \ A'_3 = \begin{pmatrix} -i & 0 \\ 0 & i \end{pmatrix}$$

Show that this is an example of the situation seen in (1) with $\lambda = \ldots$. Calculate the scalar quantities (multiple of the identity) called Casimir operators $C = A_1^2 + A_2^2 + A_3^2$ and $C' = A_1'^2 + A_2'^2 + A_3'^2$: the results should be expected from the quantum mechanical interpretation, ...

(3) Denote by $A''_1,\ A''_2,\ A''_3$ the 2×2 matrices having the same structure constants as the 3×3 matrices $A_1,\ A_2,\ A_3$, according to (1). Evaluate $\exp(aA_3)$ and $\exp(a''A''_3)$: for what values of $a \in \mathbf{R}$ one has $\exp(aA_3) = I$ and resp. for what values of a'' one has $\exp(a''A''_3) = I$?

4.29

(1) Show that all groups SU_n $(n \geq 2)$ admit a nontrivial center. Conclude: SU_n are not simple groups.

(2) Consider the group SU_2: show that its center is Z_2 (the quotient SU_2/Z_2 is isomorphic to SO_3, as well-known). The presence of not faithful representations of SU_2 is then expected: what are the representations of SU_2 which are not faithful?

4.30

(1) Consider the vectors $\mathbf{v} \in \mathbf{R}^3$ as basis space for the "basic" representation $\mathscr{R}$ by means of the 3×3 orthogonal matrices of SO_3, and then consider the tensor product $\mathbf{R}^3 \otimes \mathbf{R}^3$ (isomorphic, of course, to the 9-dimensional vector space of the real 3×3 matrices $\mathbf{M}$), together with the resulting direct product representation $\mathscr{R} \otimes \mathscr{R}$ of SO_3 acting on this space according to:

$$M_{ij} \to M'_{ij} = R_{ir} R_{js} M_{rs} = (RMR^t)_{ij}$$

(where ^{t} means matrix transposition). Show that there are invariant subspaces for this representation and obtain its decomposition into irreducible representations. According to the quantum mechanical interpretation, the vectors $\mathbf{v}$ correspond to the angular momentum $\ell = 1$; give then the interpretation of the above decomposition as a result of the combination of two angular momenta $\ell = 1$, which is often written, with evident and convenient notations, $(\ell = 1) \otimes (\ell = 1)$ or also $\underline{3} \otimes \underline{3}$.

(2) Generalize, e.g.., to tensors with 3 indices: M_{ijk}, i.e. $(\ell = 1) \otimes (\ell = 1) \otimes (\ell = 1)$ or $\underline{3} \otimes \underline{3} \otimes \underline{3}$.

4.31

(1) Consider the "basic" representation $\mathscr{R}$ of SU_3 by means of 3×3 unitary matrices U with determinant equal to 1, and the "contragredient" representation $\mathscr{R}^*$ obtained by means of matrices $U^* = (U^t)^{-1}$ where * means complex conjugation and ^{t} matrix transposition. Writing $z^i = (z_i)^*$ in such a way that upper indices are transformed by $\mathscr{R}^*$, introduce the convenient notation $\underline{3}$ and $\underline{3}^*$ to denote the basis spaces for the (inequivalent) representations[4] $\mathscr{R}$ and $\mathscr{R}^*$. Consider then the tensor products $\underline{3} \otimes \underline{3}$ and $\underline{3} \otimes \underline{3}^*$ with the corresponding transformations by means of direct products of the representations $\mathscr{R} \otimes \mathscr{R}$ and $\mathscr{R} \otimes \mathscr{R}^*$, i.e., respectively

$$T_{ij} \to T'_{ij} = U_{ir} U_{js} T_{rs} = (U T U^t)_{ij} \quad \text{and} \quad T_i^j \to T_i'^j = U_{ir} (U^*)^{js} T_r^s = (U T U^+)_i^j$$

Decompose these representations into irreducible representations of SU_3. It can be useful to introduce here the shorthand notation $T_{(\dots)}$ and $T_{[\dots]}$ to denote resp. symmetric and antisymmetric tensors, and recall that, using the totally antisymmetric tensor ε_{ijk}, the 3-dimensional representation acting on the antisymmetric tensors $T_{[ij]}$ is equivalent to the representation on the vectors $z^k = \varepsilon^{ijk} T_{[ij]}$.

(2) Decompose into irreducible representations the tensor products $\underline{3}^* \otimes \underline{3}^*$ and $\underline{3} \otimes \underline{3} \otimes \underline{3}$.

4.4 Other Relevant Applications of Symmetries to Physics

The following 4 problems give some simple examples of the application of symmetry properties to the study of differential equations. Actually, this technique can be greatly developed, introducing "less evident" symmetries of the differential equations; denoting by $u = u(x_1, x_2, \dots)$ the unknown function, one can look, e.g.., for transformations generated by infinitesimal operators of the form

$$A = \xi_1(x_1, x_2, \dots, u)\frac{\partial}{\partial x_1} + \xi_2(x_1, x_2, \dots, u)\frac{\partial}{\partial x_2} + \dots + \varphi(x_1, x_2, \dots, u)\frac{\partial}{\partial u}$$

where the functions ξ_i and φ can be arbitrary functions (with some obvious regularity assumptions) and not only constants or linear functions of the independent variables x_i as in usual cases, and in all the examples considered in this book. In addition, also the dependent variable u is assumed to be subjected to a transformation, as shown by the presence of the term φ in the expression

---

[4] It can be useful to point out that, differently from all the groups SU_n with $n > 2$, the "basic" irreducible representations $\mathscr{R}$ and $\mathscr{R}^*$ of SU_2, by means of 2×2 unitary matrices, are equivalent.

of the generator A. A presentation of these methods and of their applications is given, e.g., in the book by P.J. Olver, see Bibliography.

4.32
The Laplace equation in $\mathbf{R}^2$

$$u_{xx} + u_{yy} = 0\,, \qquad u = u(x, y)$$

is clearly symmetric (or, more correctly, invariant[5]) under the rotation group O_2 in the plane x, y. This implies that if $u = u(x, y)$ is a solution to this equation, then also $\widetilde{u} = u(x', y')$, with $x' = x\cos\varphi - y\sin\varphi$, etc. is a solution for any φ. Observing that, for instance, $u = \exp x \cos y = \mathrm{Re}\,(\exp z)$, with $z = x + iy$, solves the equation, construct a family of other solutions of the equation. Verify that this new family of solutions is the same one can obtain by means of the transformation $z \to z' = \exp(i\varphi)z$. Using the same procedure, construct other families of solutions starting from some known one.

4.33
The non-homogeneous Laplace equation for $u = u(x, y)$

$$u_{xx} + u_{yy} = r^n\,, \qquad n = 0, 1, \ldots$$

with $r^2 = x^2 + y^2$, is clearly symmetric under the rotation group O_2 in the plane x, y. One then can look for the existence of solutions which are *invariant* under O_2, i.e., for solutions of the form $u = f(r)$. Recalling the well-known expression of the Laplace equation in terms of the polar coordinates r, φ, transform the given PDE into an ODE for the unknown $f(r)$ and solve this equation.

4.34
(1) The heat equation

$$u_t = u_{xx}\,, \qquad u = u(x, t)$$

is symmetric under independent translations of x and t, and therefore under any combination of these translations. Look then for solutions which are functions only of $x - vt$ ("traveling wave solutions" with arbitrary velocity v). Transform the PDE into an ODE for the unknown function $u = f(s)$ of the independent variable $s = x - vt$ and obtain the most general solution of this form.

(2) Do the same for the d'Alembert equation

$$u_{xx} = u_{tt}\,, \qquad u = u(x, t)$$

What traveling waves are obtained?

---

[5]the equation indeed is left *invariant* under the transformations of the group, which is called the *symmetry group of the equation*; symmetric and invariant are synonymous in this context.

4.35
(1) The d'Alembert equation

$$u_{xx} = u_{tt} \,, \qquad u = u(x, t)$$

is symmetric under the Lorentz transformations. Starting, for instance, with the solution $u = x^3 + 3xt^2$, construct a family of other solutions.

(2) Consider the non-homogeneous d'Alembert equation

$$u_{xx} - u_{tt} = (x^2 - t^2)^n \,, \qquad n = 0, 1, \ldots$$

Observing that the equation is symmetric under Lorentz transformations, one can look for the existence of solutions which are Lorentz-invariant, i.e., for solutions of the form $u = f(s)$ where $s = x^2 - t^2$. Transform then the given equation into an ODE for the unknown $f(s)$ and solve this equation.

4.36
Consider a quantum mechanical system exhibiting spherical symmetry, e.g., a particle placed in a spherically symmetric (radial) potential $V(r)$. The eigenvalues of its Hamiltonian (Schrödinger equation), i.e., the energy levels, are expected to have a degeneracy ..., as a consequence of the symmetry under the group SO_3 and the Schur lemma. If the system is placed into a (weak) uniform magnetic field, the SO_3 symmetry is broken and the surviving symmetry is SO_2 (*not* O_2: way?). What consequence can be expected about the degeneracy of the energy levels? What changes if instead the system is placed into an uniform electric field, where the residual symmetry is O_2? This is a qualitative description, based only on symmetry arguments and in particular on Schur lemma, of the Zeeman and respectively the Stark effects.

In the following problems, which deal with the symmetry properties of the hydrogen atom and the harmonic oscillators in quantum mechanics, it is convenient to redefine the generators of the Lie algebras in order to have Hermitian *operators, introducing now a factor* i in the definitions used in all previous problems (see, e.g.., Problems 4.13, 4.18). *Accordingly, for instance, the generator* A_3 *of* SO_3 *(with clear notation) will be, in its basic* 3×3 *matrix form,*

$$A_3 = \begin{pmatrix} 0 & -i & 0 \\ i & 0 & 0 \\ 0 & 0 & 0 \end{pmatrix}$$

with real eigenvalues ± 1 and 0, *or, in differential form,*

$$A_3 = i\Big(y\frac{\partial}{\partial x} - x\frac{\partial}{\partial y}\Big) = -i\frac{\partial}{\partial \varphi}$$

4.37

In view of the application to the hydrogen atom in the next problem, this one is devoted to the study of the group SO_4 and its algebra. Consider the space $\mathbf{R}^4$ with cartesian coordinates x_1, x_2, x_3, x_4.

(1) Show that SO_4 has 6 parameters and then its algebra is 6-dimensional. Denote by A_1, A_2, A_3 the Hermitian generators of the subgroup SO_3 of the rotations in the subspace $\mathbf{R}^3$ with basis x_1, x_2, x_3, and by B_1, B_2, B_3 the generators of the rotations in the subspaces (x_1, x_4), (x_2, x_4), (x_3, x_4). Either using a 4×4 matrix representation or the differential representation, show that the algebra of SO_4 is described by the following commutation rules, with $i, j = 1, 2, 3$,

$$[\, A_i\,,\, A_j\,] = i\,\varepsilon_{ijk}A_k\,, \quad [\, B_i,\, B_j\,] = i\,\varepsilon_{ijk}A_k\,, \quad [\, A_i\,,\, B_j\,] = i\,\varepsilon_{ijk}B_k$$

Compare with the Lie algebra of the Lorentz group $\mathscr{L}$ (Problem 4.12, q.(2)): there is only an (essential!) change in a sign.

(2) Put

$$M_i = \frac{1}{2}(A_i + B_i)\,, \quad N_i = \frac{1}{2}(A_i - B_i)$$

and show that the operators M_i, N_i satisfy

$$[\, M_i\,,\, M_j\,] = i\,\varepsilon_{ijk}M_k\,, \quad [\, N_i\,,\, N_j\,] = i\,\varepsilon_{ijk}N_k\,, \quad [\, M_i\,,\, N_j\,] = 0$$

so that the Lie algebra of SO_4 (*not* the group, which is indeed a simple group!) is isomorphic to the algebra of the direct product $SU_2 \times SU_2$.

(3) Recalling that the irreducible representations of SU_2 are specified by an integer or half-integer number $j = 0,\ 1/2,\ 1, \ldots$, and have dimension $N = 2j + 1 = 1, 2, \ldots$, describe the irreducible representations of $SU_2 \times SU_2$ and find their dimensions.

(4) Show that SO_4 has rank 2; verify then that the two Casimir operators (scalar quantities, multiple of the identity)

$$C_1 = \mathbf{A}^2 + \mathbf{B}^2 = 2\big(\mathbf{M}^2 + \mathbf{N}^2\big) \quad \text{and} \quad C_2 = \mathbf{A}\cdot\mathbf{B} = \mathbf{M}^2 - \mathbf{N}^2$$

commute with the algebra of SO_4 and find the values of C_1 and C_2 in the various representations.

4.38
Consider the Schrödinger equation for the hydrogen atom, with usual notations,

$$\mathrm{H}u \equiv -\frac{\hbar^2}{2m}\Delta u - \frac{e^2}{r}u = E u$$

As any other Hamiltonian exhibiting spherical symmetry, this Hamiltonian commutes with the 3 operators A_i, where e.g.. $A_3 = i(y\,\partial/\partial x - x\,\partial/\partial y) = xp_y - yp_x$, etc., proportional to the operators $L_i = \hbar A_i$, describing, as well-known, the angular momentum:

$$[\mathrm{H}, L_i] = 0$$

It is known from classical mechanics that the 3 components of the Runge-Lenz vector

$$\widetilde{\mathbf{R}} = \mathbf{p}\times\mathbf{L} - me^2\frac{\mathbf{r}}{r}$$

are constants of motion for the Kepler problem. This has a precise counterpart in quantum mechanics: putting indeed, in a correct quantum mechanical form,

$$\mathbf{R} = \frac{1}{2}\big(\mathbf{p}\times\mathbf{L} - \mathbf{L}\times\mathbf{p}\big) - me^2\frac{\mathbf{r}}{r}$$

one can find

$$[\mathrm{H}, R_i] = 0$$

Introducing the dimensionless Hermitian operators

$$B_i = \frac{1}{\hbar\sqrt{-2mH}}R_i\,, \qquad i = 1, 2, 3$$

(only bound states will be considered, then the energy is negative) it can be shown, after some tedious (not requested) calculus, that the 6 operators $A_i,\ B_i$ satisfy precisely the commutation rules of the algebra of SO_4, isomorphic to the algebra $SU_2 \times SU_2$ (obtained in the previous problem).

(1) With the notations of the previous problem, i.e., $\mathbf{M} = (\mathbf{A}+\mathbf{B})/2\,, \mathbf{N} = (\mathbf{A} - \mathbf{B})/2$, show that

$$\mathbf{L}\cdot\mathbf{R} = \mathbf{A}\cdot\mathbf{B} = \mathbf{M}^2 - \mathbf{N}^2 = 0$$

Conclude from the last equality: what representations of $SU_2 \times SU_2$ are involved in the hydrogen atom? and what degeneracy can be expected for the eigenvalues of the energy of the hydrogen atom?

(2) It can be also shown that

$$(\mathbf{A}^2 + \mathbf{B}^2 + 1)\,H = -\frac{me^4}{2\hbar^2}$$

Observing that $\mathbf{A}^2 + \mathbf{B}^2 = 4\mathbf{M}^2$ and recalling that the eigenvalues of $\mathbf{M}^2$ are ..., deduce the eigenvalues of the energy of the hydrogen atom.

(3) Decompose the n^2-dimensional irreducible representation corresponding to the n-th energy level as a sum of the irreducible representations of the rotation subgroup SO_3 (i.e., find the angular momenta ℓ contained in the n-th level); to this aim, observe that $\mathbf{L} = \hbar(\mathbf{M} + \mathbf{N})$, therefore, the angular momentum ℓ can be obtained as a superposition of the two "spin" j_M and $j_N = j_M$, and therefore $\ell = 0, 1, \dots, 2j_M = \ell_{Max} = n - 1$. Show that

$$\sum_{\ell=0}^{\ell_{Max}} (2\ell + 1) = n^2\,, \qquad n = 1, 2, \dots$$

4.39

As in the case of the hydrogen atom, also the 3-dimensional harmonic isotropic oscillator admits, besides the SO_3 rotational symmetry, an additional symmetry.

(1) Consider first the 3-dimensional harmonic isotropic equation in classical mechanics: its Hamiltonian is (with $m = k = 1$ and with standard notations)

$$H = \frac{1}{2}\mathbf{p}^2 + \frac{1}{2}\mathbf{r}^2$$

Put

$$\zeta_j = \frac{1}{\sqrt{2}}(x_j + ip_j)\,, \qquad j = 1, 2, 3$$

and show that H can be written as $H = (\boldsymbol{\zeta}, \boldsymbol{\zeta})$, i.e., as a scalar product in $\mathbf{C}^3$. Show also that the 9 quantities $\zeta_j^* \zeta_k$ are constants of motion: $(d/dt)\zeta_j^*\zeta_k = 0$.

(2) *(a)* Write the Schrödinger equation for the oscillator (with $m = k = 1$)

$$\mathrm{H}\, u \equiv -\frac{\hbar^2}{2}\Delta\, u + \frac{1}{2}\mathbf{r}^2\, u = E\, u$$

Put now (with $\hbar = 1$)

$$\eta_j = \frac{1}{\sqrt{2}}\left(x_j + \frac{\partial}{\partial x_j}\right)\,, \quad \eta_j^+ = \frac{1}{\sqrt{2}}\left(x_j - \frac{\partial}{\partial x_j}\right)\,, \qquad j = 1, 2, 3$$

show that

$$\left[\eta_j\,, \frac{\partial}{\partial x_k}\right] = -\frac{1}{\sqrt{2}}\delta_{jk} \;\text{ and }\; [\eta_j^+\,, \eta_k] = -\delta_{jk}$$

and that the 9 operators

$$A_k^j = \eta_j^+ \eta_k\,, \qquad j, k = 1, 2, 3$$

commute with the Hamiltonian operator H.

(*b*) Verify that the following Hermitian operators

$$i(A^2_1 - A^1_2)\,,\ i(A^3_2 - A^2_3)\,,\ i(A^1_3 - A^3_1)\,,$$

$$A^1_2 + A^2_1\,,\ A^2_3 + A^3_2\,,\ A^3_1 + A^1_3\,,\ A^1_1 - A^2_2\,,\ A^1_1 + A^2_2 - 2A^3_3\,,\ A^1_1 + A^2_2 + A^3_3$$

generate the Lie algebra of the group $U_3 = SU_3 \times U_1$. Verify that the first 3 operators are just the generators of the rotation subgroup SO_3, that the last operator, i.e., $\mathtt{Tr}\,\mathbf{A}$, generates U_1 and satisfies $\mathtt{Tr}\,\mathbf{A} = (1/2)(-\Delta + r^2 - 3) = \mathtt{H} - (3/2)$.

(*c*) For completeness (this is indeed a well-known procedure), find the eigenvectors and eigenvalues of $\mathtt{H}$: to this end, find first the eigenfunction $v_0(x_1)$ of η_1 with zero eigenvalue: $\eta_1\, v_0(x_1) = 0$: therefore $A^1_1\, v_0(x_1) = \eta^+_1 \eta_1\, v_0(x) = 0$. Verify that

$$[A^1_1\,,\ \eta^+_1] = \eta^+_1$$

put now (this is an example of *ladder operator*, see Problem 1.44, q. (2))

$$v_1 = v_1(x_1) = \eta^+_1\, v_0(x_1) \quad ; \quad v_2 = v_2(x_1) = \eta^+_1\, v_1(x_1) \quad ; \quad \text{etc.}$$

and deduce $A^1_1\, v_1 = \ldots,\ A^1_1\, v_2 = \ldots$, and so on; do the same for A^2_2 and A^3_3. Conclude: what are the eigenvalues of $\mathtt{Tr}\,\mathbf{A}$? Show that the fundamental state of the 3-dimensional oscillator is given by $u_0 = u_0(x_1, x_2, x_3) = v_0(x_1)v_0(x_2)v_0(x_3) \propto \exp(-r^2/2)$ with energy $E = 3/2$ (in unit $\hbar = 1$, and $m = k = 1$); construct next the 3 states at the first excited energy level $E = 5/2$ (these correspond to the representation $\underline{3}$ of U_3), and so on for higher energy levels.

(*d*) Show that the irreducible representations of U_3 involved in the energy eigenstates of the harmonic oscillator are those having the symmetric tensors $T_{(i_1, i_2, \ldots)}$ as basis space (see Problem 4.31 for what concerns the irreducible representations of SU_3 and the notations). *Hint*: consider for instance the second excited state, which is given by $u \propto x_j x_k \exp(-r^2/2)$, with degeneracy 6, …, the third excited state with degeneracy 10, ….

(*e*) Decompose the 6-dimensional representation considered in the previous question (*c*) as a sum of irreducible representations of SO_3 (i.e., find the angular momenta ℓ contained in the second excited level). Generalize to the other levels.

(*f*) Generalize to the N-dimensional isotropic oscillators with generic N.

Answers and Solutions

Problems of Chap. 1

(1.1)
(2) E.g., choosing $c_n = 1/n$ one has $\int_{-\infty}^{+\infty} f_n(x)\,dx = 1$; choosing $c_n = 1/\sqrt{n}$ the integral $\to \infty$. In both cases, $\sup_{x\in\mathbf{R}} |f_n(x)| = |c_n| \to 0$
(3) *(a)* If $c_n = 1/n$ one has $\|f_n\|_{L^2} = 1/\sqrt{n} \to 0$, but $\|f_n\|_{L^1} = 1$
(b) Let c_n be such that $\|f_n\|_{L^1} = n\,|c_n| \to 0$; on the other hand, $\|f_n\|_{L^2} = \sqrt{n}|c_n| < n|c_n| \to 0$; then, for the given functions, $f_n(x) \to 0$ in $L^1(\mathbf{R})$ implies $\to 0$ in $L^2(\mathbf{R})$
(4) *(a)* $f_n(x)$ does not converge at the point $x = 0$
(b) Let c_n be such that $\|f_n\|_{L^2} = |c_n|/\sqrt{n} \to 0$; on the other hand $\|f_n\|_{L^1} = |c_n|/n < |c_n|/\sqrt{n} \to 0$; then, for the given functions, $f_n(x) \to 0$ in $L^2(\mathbf{R})$ implies $\to 0$ in $L^1(\mathbf{R})$
(c) If $c_n = \sqrt{n}$ then $\|f_n\|_{L^1} = 1/\sqrt{n} \to 0$ but $\|f_n\|_{L^2} = 1$.

(1.2)
(1) If $f(x) \in L^2(I)$, then

$$\|f\|_{L^1} = \int_I |f|dx = (\chi_I, |f|) \le \sqrt{\mu(I)}\|f\|_{L^2}$$

where $\chi_I(x)$ is the characteristic function of the interval I. Conversely, e.g.. $f(x) = 1/\sqrt{x} \in L^1(0,1)$ but $\notin L^2(0,1)$
(2) E.g., $f(x) = 1/\sqrt{x}$ for $0 < x < 1$ (and $= 0$ elsewhere) $\in L^1(\mathbf{R})$ but $\notin L^2(\mathbf{R})$.
E.g., $f(x) = x/(1+x^2) \in L^2(\mathbf{R})$ but $\notin L^1(\mathbf{R})$
(3) No: $\|f\|_{L^2} = \left(\int_{-1}^{+1} |f(x)|^2 dx\right)^{1/2} \le 2\left(\sup_{-1\le x\le 1} |f(x)|^2\right)^{1/2} \le 2\varepsilon$.
Yes: choose $f(x) = 1$ in an interval of length ε^2 and $= 0$ elsewhere

G. Cicogna, *Exercises and Problems in Mathematical Methods of Physics*, Undergraduate Lecture Notes in Physics,
https://doi.org/10.1007/978-3-030-59472-5

(4) Yes: choose $f(x) = \varepsilon$ in an interval of length $1/\varepsilon^2$ and $= 0$ elsewhere.
Yes: the same example as in (3)

(1.3)
(1) Let $f_n(x) \in L^2(\mathbf{R})$:

$$\int_{-\infty}^{+\infty} |f_n(x)|\,dx = \int_{-n}^{+n} |f(x)|\,dx = (\chi_n, |f|) \le \sqrt{2n}\|f\|_{L^2}$$

where $\chi_n = \chi_n(x)$ is ..., then $f_n(x) \in L^1(\mathbf{R})$. On the other hand, $|f_n(x)| \le |f(x)|$ and $\int_{-\infty}^{+\infty}|f_n(x)|^2\,dx \le \int_{-\infty}^{+\infty}|f(x)|^2\,dx$ then $f_n(x) \in L^2(\mathbf{R})$. Finally,

$$\|f(x) - f_n(x)\|_{L^2} = \int_{|x|\ge n} |f(x)|^2 dx \to 0$$

This means that the set of truncated functions is *dense* in $L^2(\mathbf{R})$, as well as the set of the functions $f \in L^1(\mathbf{R}) \cap L^2(\mathbf{R})$. The same holds for the functions with compact support. Their *closure* is the space $L^2(\mathbf{R})$; as a consequence, they are not Hilbert subspaces
(2) Yes, as in (1): recall, e.g.., that the Hermite functions $u_n(x) = \exp(-x^2/2)\,H_n(x)$, where $H_n(x)$ are polynomials of degree $n = 0, 1, 2, \ldots$, are a complete set in $L^2(\mathbf{R})$

(1.4)
(2) Yes: if $f(x) \in L^1(\mathbf{R}) \cap L^2(\mathbf{R})$, the suggested procedure is enough; if $f(x) \in L^2(\mathbf{R})$ but $\notin L^1(\mathbf{R})$, then first approximate $f(x)$ with a function $g(x) \in L^1(\mathbf{R}) \cap L^2(\mathbf{R})$, etc.

(1.5)
(1) This is a closed (and therefore a *Hilbert*) subspace: indeed, let $f_n(x) \in V$ be a Cauchy sequence and let $f(x) = \lim_{n\to+\infty} f_n(x) \in L^2(-a, a)$ $\big(f(x)$ certainly exists because $L^2(-a, a)$ is a Hilbert space, closed by definition$\big)$. Thanks to the *continuity of the scalar product*, $\int_{-a}^{+a} f_n(x)\,dx \equiv (\chi_a, f_n) = 0$ implies also $(\chi_a, f) = 0$, then $f(x) \in V$, and V is closed. More explicitly: $|(\chi_a, f)| = |(\chi_a, f - f_n)| \le \|\chi_a\|\,\|f - f_n\| \to 0$. The orthogonal complement is the 1-dimensional subspace of the constant functions in the given interval
(2) The subspace of zero mean-valued functions is *dense* in $L^2(\mathbf{R})$: see previous problem

(1.6)
(1) $\|g_n - u\|_{L^2} = 2\sqrt{1/3n} \to 0$
(2) This should be intuitively obvious, recalling that the approximation of a function $f(x) \in L^2$ with a function $g(x)$ belonging to these subspaces is to be intended *not pointwise, but in L^2-norm, i.e., in mean-square*. It is clear that the procedure used for (1) can be easily extended to any $f(x) \in L^2(I)$, etc.

Another argument: one can observe that, given any orthogonal complete set $\{e_n(x)\}$ in $L^2(I)$ of C^∞ functions, then (it is not restrictive to choose $x_0 = 0$ and an interval I containing $x_0 = 0$) also $u_n(x) = x\, e_n(x)$ is a complete system in $L^2(I)$, and that the same is true for $v_n(x) = x^k e_n(x)$ for any integer k, and also, e.g., for $w_n(x) = \exp(-1/x^2)e_n(x)$ $\big($cf. Problem 1.14, q. (2)$\big)$. Recall then that the finite linear combinations of these complete systems provide dense subspaces with the required properties

(1.7)
(1) Yes: $\lim \ldots \le \|f\|/\sqrt{2N} \to 0,\ \forall f(x)$
(2) This limit is trivially zero for even functions, and is zero in the dense set of functions with compact support; it is equal to 1 if, e.g., $f(x) = x/(1+x^2)$, is $+\infty$ if, e.g., $f(x) = 1/x^{2/3}$ for $x > 1$ (and $= 0$ for $x < 1$); in the last case the integral $\int_{-N}^{+N} \ldots$ behaves for large N as $N\cos(N^{1/3})$

(1.8)
(1–2) With, e.g., $\delta_n = 1/|n|^3$ $(n \neq 0)$, the function $\in L^1(\mathbf{R})$ and also $\in L^2(\mathbf{R})$.
A function equal to n on the intervals I_n and $\delta_n = 1/n^4$ (resp. $\delta_n = 1/|n|^5$) is an example of unbounded function $\in L^1(\mathbf{R})$ $\big($resp. $\in L^2(\mathbf{R})\big)$

(1.9)
(1) $\|e_n - e_m\| = \sqrt{2}$ (if $n \neq m$, of course) ; $(e_n, x) \to 0,\ \forall x \in H$, thanks to the Bessel inequality: $\|x\|^2 \ge \sum_n |(e_n, x)|^2$, then $\{e_n\}$ converges weakly to zero
(2) *(a)* $\|x_n - x\|^2 = \|x_n\|^2 - (x_n, x) - (x, x_n) + \|x\|^2 \to 0$
(b) $\|x_n\| \le \|x_n - x\| + \|x\| \le \varepsilon + M_1$ with clear notations

(1.10)
V_1 is not closed; its closure $\overline{V_1}$ is the Hilbert subspace $L^2_{even}(-1, 1)$ of the *even* functions in the interval $(-1, 1)$; the orthogonal complementary subspace is the Hilbert space $L^2_{odd}(-1, 1)$ of the *odd* functions.
V_2: the same as V_1.
V_3 is a Hilbert subspace $\big($see Problem 1.5, q. (1)$\big)$; the orthogonal complementary subspace is the space $L^2(-1, 0) \oplus \mathbf{C}(0, 1)$ where $\mathbf{C}(0, 1)$ is the 1-dimensional space of the constants functions in the interval $(0, 1)$.
V_4 is dense in $L^2(-1, 1)$ $\big($see Problem 1.6, q.(2)$\big)$, i.e., $\overline{V_4} = L^2(-1, 1)$.

(1.11)
(1) The set of the polynomials is not a Hilbert space; its *closure* is just $L^2(-1, 1)$
(2) $\{x^{2n}\}$ is clearly a complete set in the Hilbert space $L^2_{even}(-1, 1)$ of the *even* functions; then given any $f(x) \in L^2(0, 1)$, it is enough to refer to its *even* prolongation to the full interval $(-1, 1)$. The similar conclusion holds for $\{x^{2n+1}\}$ and the *odd* prolongation of any given function $f(x) \in L^2(0, 1)$
(3) Look for a function $g(x)$ such that $\big(x^{N+m}, g(x)\big) = 0,\ \forall m \ge 0$: one has $0 = \big(x^{N+m}, g(x)\big) = \big(x^m, x^N g(x)\big),\ \forall m \ge 0$, then $x^N g(x) = 0$ and $g(x) = 0$ $\big($possibly

apart from the point $x = 0$, but this is not relevant, see Problems 1.6, q.(2) and 1.5, q.(2))

(1.12)
(1) Yes: look for $z \in H$ such that $(v_n, z) = 0$, $\forall n = 2, 3, \dots$: this implies $(e_n, z) = (e_1, z)$, $\forall n$, then necessarily $z = 0$, recalling that $\|z\|^2 = \sum_n |(e_n, z)|^2$
(2) Let $z \in H$ be such that $(v_n, z) = 0$, $\forall n$, then $(v_n, z) = (e_n, z) - (w, z) = 0$ or $(e_n, z) = \text{const}$. This constant must be zero (see the argument for q. (1)), then $z = 0$, which implies that the set v_n is complete
(3) $\alpha_n \in \ell^2$
(4) Let $z \in H$ be such that $(v_n, z) = 0$, i.e. $\alpha\, z_n = \beta\, z_{n+1}$ where $z_n = (e_n, z)$, and $|z_{n+1}| = |\alpha/\beta||z_n|$. Then, if $|\alpha| < |\beta|$ there is a nonzero vector $z \in H$ and the set is *not* complete; if $|\alpha| \geq |\beta|$ then z must be zero and the set is complete

(1.13)
(1) Yes. Yes. Use an argument as in the above problem, but now $n \in \mathbf{Z}$ and this makes the difference from q. (4) of the above problem
(2) $v_n = e_n(x)(1 - e^{ix})$ and $w_n = e_n(x)(\alpha - \beta e^{ix})$ are complete sets in $L^2(0, 2\pi)$ (see Problem 1.14, q. (2))

(1.14)
(1) *(a)* and *(b)* are complete (for any $P(x)$): use the result of q.(2)
(c) is not complete: the constant function 1 is the only even function in the set.
(d), *(e)*, *(g)*, *(h)*: the only functions $\in L^2(-\pi, \pi)$ orthogonal to $\{\cos nx,\ \sin nx\}$ are the constants. Then: *(d)* is not complete, because the function $g(x) = 1$ is orthogonal to all the functions of the set. *(e)* is complete, because $(x^2, 1) \neq 0$ then there is no function orthogonal. For the case *(g)*, let $g = g(x)$ be a function such that $\big(x \cos nx, g(x)\big) = \big(\cos nx, x\, g(x)\big) = 0$ and $\big(x \sin nx, g(x)\big) = \big(\sin nx, x\, g(x)\big) = 0$, then $x\, g(x) = \text{const}$, but $g(x) = \text{const}/x \notin L^2(-\pi, \pi)$, then the set is complete. In the case *(h)*, the same argument gives $g(x) = \text{const}/x^{1/3} \in L^2(-\pi, \pi)$, and the set is not complete
(f) is complete: let $g = g(x) \in L^2(\pi, \pi)$ be orthogonal to this set. $\{\sin nx\}$ is complete in the subspace $L^2(0, \pi)$ then $\{\sin\, n\, |x|\}$ is complete in the subspace of even functions in $L^2(-\pi, \pi)$; therefore $g(x)$ must be odd. On the other hand, imposing $\big(x, g(x)\big) = \big(1, x\, g(x)\big) = 0$ and $\big(x \cos nx, g(x)\big) = \big(\cos nx, x\, g(x)\big) = 0$ one gets $x\, g(x) = 0$, or equivalently $g(x) = 0$, for the completeness of $\{1,\ \cos nx\}$ in the set of even functions in $L^2(-\pi, \pi)$ (clearly, $x\, g(x)$ is even).
(2) Apart from the obvious restriction $h(x)e_n(x) \in L^2(I)$, the condition concerns only the zeroes of $h(x)$: this function must have at most isolated zeroes. Indeed, imposing that a function $g(x) \in L^2(I)$ is such that $\big(h(x)\, e_n(x), g(x)\big) = \big(e_n(x), h^*(x)\, g(x)\big) = 0$, $\forall n$, gives $h^*(x) g(x) = 0$ and then $g(x) = 0$ "almost everywhere" (i.e., $g(x)$ is *equivalent* to the zero function in $L^2(I)$)

(1.15)
(1) Certainly, $|a_n| \to 0$; on the other hand, when $|a_n| < 1$ then $|a_n|^2 < |a_n|$, ... Conversely, with $a_n = 1/n$ one obtains a sequence $\in \ell^2$ but $\notin \ell^1$

(2) ℓ^1 certainly contains all the *finite* combinations of the "canonical" orthonormal complete set $e_1 = (1, 0, 0, \ldots)$, $e_2 = (0, 1, 0, 0, \ldots)$, then ...

(1.16)
(1) Imposing $(w_n, z) = 0\,\forall n$ with $z \in \ell^2$ implies $z_1 = z_2 = \ldots = z_n = \ldots,\ \forall n$, then $z = 0$ (see Problem 1.12, q.(1))
(2) $\ell^{(0)}$ contains all the *finite* combinations of the vectors w_n, then ...
(3) $\|z_n - e_1\|^2 = 1/n \to 0$

(1.17)
(1) Only (γ) is correct; (β) is wrong because $\sin 2\pi x \notin L^2(0, \infty)$
(2) No: e.g.., $g(x) = \sin 2x$ for $0 \le x \le \pi$ and $= 0$ elsewhere is orthogonal to all the functions of this set
(3) *(a)* Yes: recall that $\{\sin nx\}$ is an orthogonal set in the space $L^2(0, \pi)$
(b) The set is not complete: e.g., $f(x) = \sin x$ for $\pi \le x \le 2\pi$ and $= 0$ elsewhere is orthogonal to all v_n

(1.18)
(1) Yes, clearly, for the first question. No for the second: the function $(\sin x)/x \in L^2(0, \pi)$ is orthogonal to the subset $\{x \sin nx\}$ if $n = 2, 3, \ldots$
(2) Yes. Yes: a function orthogonal to the set $\{x^2 \sin nx\}$ with $n = 2, 3, \ldots$ should be (proportional to) $(\sin x)/x^2$ which is not in $L^2(0, \pi)$

(1.19)
(a) Yes: impose the condition $0 = \big(\exp(-nx), f(x)\big) = \int_0^{+\infty} \ldots = \int_0^1 y^{n-1} f(\cdot)\, dy$ and recall the completeness of y^{n-1} in $L^2(0, 1)$
(b) No: e.g.., the function $g(x) = -1$ for $-2\pi < x < 0$ and $= 1$ for $0 < x < 2\pi$ is orthogonal to all the functions of this set
(c) Yes: recall that the set is complete in $L^2(0, 2\pi)$ ...
(d) No: e.g.., the function $g(x) = \sin x \sin 2y$ is orthogonal to all the functions of this set.
(e) No: e.g.., the function $g(x, y) = g_1(x) \sin y$, where $g_1(x)$ is any function orthogonal to $\exp(-x)$ in $L^2(0, +\infty)$, is orthogonal to all the functions of this set
(f) No: e.g.., the function

$$g(x) = \begin{cases} -\exp(x^2) \text{ for } -2\pi < x < 0 \\ \exp(x^2) \quad \text{for} \quad 0 < x < 2\pi \end{cases}$$

is orthogonal to all the functions in the set

(1.20)
(1) $f_1(x) = (4/\pi)\big(\sin x + \frac{1}{3}\sin 3x + \frac{1}{5}\sin 5x) + \frac{1}{7}\sin 7x + \ldots\big)$
For the convergence of the Fourier expansion of $f_1(x)$, see q. (3)
$f_2(x) = \pi/2 - (4/\pi)\big(\cos x + \frac{1}{9}\cos 3x + \frac{1}{25}\cos 5x + \ldots\big).$

The expansion of $f_2(x)$ converges *uniformly* to $f_2(x)$
(2) The series (a) converges to $\pi/4$, the series (b) to $\pi^2/8$
(3) The series converges pointwise to the square-wave

$$S(x) = \begin{cases} 1 & \text{for } \ 0 < x < \pi \\ -1 & \text{for } \ -\pi < x < 0 \end{cases}$$

with its periodic prolongation to all $x \in \mathbf{R}$ with period 2π.
At $x = 0, \ \pm\pi$ the series converges to zero, at $x = 3\pi/2$ converges to -1

(1.21)
(1–2) Writing

$$f_1(x) = \sum_{n=1}^{\infty} a_n^{(1)} \sin nx \quad , \quad f_2(x) = \sum_{n=1}^{\infty} a_n^{(2)} \sin nx$$

one finds $a_n^{(1)} = (-1)^{n+1}/n$ and $a_n^{(2)} = -1/n$

(1.22)
If $f(x) = x \in L^2(0, a)$, then $\widetilde{f_1}(x) = x$ in $0 < x < a$, $\widetilde{f_1}(0) = f_1(a) = a/2$, and $\widetilde{f_1}(x)$ is periodic with period a.
$\widetilde{f_2}(x) = |x|$ in $-a \le x \le a$ and is periodic with period $2a$; in this case the series converges uniformly for all $x \in \mathbf{R}$.
$\widetilde{f_3}(x) = x$ in $-a < x < a$, $\widetilde{f_3}(\pm a) = 0$; $\widetilde{f_3}(x)$ is periodic with period $2a$

(1.23)
(1) $\widetilde{f}(x, y) = 0$ along the lines $x = n\pi$, $y = m\pi$; $n, m \in \mathbf{Z}$; $\widetilde{f}(x, y) = 1$ in the interior of Q; $\widetilde{f}(x, y) = -1$ in the interior of the 4 squares adjacent to Q; $\widetilde{f}(x, y) = 1$ in the other 4 squares near Q, etc.
(2) $\widetilde{f}(x, y) = \begin{cases} \sin x & \forall x \in \mathbf{R}, \ 0 < y < \pi \\ -\sin x & \forall x \in \mathbf{R}, \ \pi < y < 2\pi \end{cases}$ etc

(1.24)
(1) The condition $\sum_n |a_n| < \infty$ ensures that a series of the form $\sum_n a_n \exp(inx)$ is uniformly convergent, and then converges to a continuous function, including its periodic prolongation out of the interval $(0, 2\pi)$
(2) $f(x)$ is at least h times continuously differentiable
(4) $f(x) \in C^\infty$, because $\{n^k/2^{|n|}\} \in \ell^1$ for any $k = 0, 1, 2, \ldots$

(1.25)
(1) The function is even, continuously differentiable, its second derivative $f''(x) \in L^2(-\pi, \pi)$, but is expected to be not continuous
(2) Recall that a series of the form $\sum_n \alpha_n \sin nx$ is uniformly convergent (and then converges to a continuous function, with its periodic prolongation out of the interval

$(-\pi,\pi)\big)$ if $\sum_n |\alpha_n| < \infty$, see the above problem, q.(1). This condition is satisfied in this case, indeed we have $\alpha_n = a_n/n$ where $a_n \in \ell^2$, and $\sum_n |a_n|/n$ can be viewed as the scalar product between two sequences $\in \ell^2$

(1.26)
(1) *(a)* $\pi\sqrt{\pi}/2$ and resp. 0 (remember that the Fourier expansion produces a periodic prolongation of the function ...)
(b) No, the periodic prolongation of $f(x)$ is discontinuous, see Problem 1.24, q.(1)
(2) No, $f'(x) \notin L^2(-\pi,\pi)$
(3) Yes the first question; no the second question, because the periodic prolongation of $f(x)$ is differentiable in $L^2(-\pi,\pi)$, but not continuously, see Problem 1.24, q.(2)

(1.27)
(1) The Fourier expansion converges just to the function $f_1(x)$; notice indeed that the functions v_{2m+1} are *even* with respect to the point $x = \pi/2$
(2) The Fourier expansion gives zero, $f_2(x)$ is *odd* with respect to the point $x = \pi/2$
(3) The expansion converges to the constant function $\pi/2$; notice that $f_3(x) = \pi/2 + f_2(x)$
(4) Yes: extend $\big($using the parity property of the set $\{v_{2m+1}\}\big)$ to the interval $(0,\pi)$

(1.28)
The set contains only one *even* function: the constant, then ... E.g., the Fourier expansion of $f_3(x)$ converges to the constant function $g(x) = 1/2$ in $(-\pi,\pi)$

(1.29)
The expansion of f_1 converges to the constant function $= 1/2$ in $0 < x < 4\pi$. The expansion of f_2 converges just to f_2

(1.30)
The expansion of $f_1(x)$ converges to $f_1(x)$. The expansion of $f_2(x)$ converges to the function

$$g(x) = \begin{cases} (4/\pi)\sin x & \text{for} \quad 0 < x < \pi \\ 0 \quad \text{for} \quad x > \pi \end{cases}$$

indeed, $\forall n > 1$ the functions $\sin nx$ in $(0, 2\pi)$ are orthogonal to the constant

(1.31)
(1) The Fourier expansion is a piecewise constant function; in each interval $(n-1, n)$ this function takes the value $c_n = \int_{n-1}^{n} f(x)\,dx$, the mean value of the function $f(x)$ in that interval.
(2) Pointwise, not uniform convergence to 0. No Cauchy property in the $L^2(0,\infty)$ norm, indeed $\|u_n - u_m\|^2 = 2$ (if $n \neq m$, of course). Weak L^2-convergence to 0: indeed, due to the Bessel inequality $\sum_n |(u_n, g)|^2 \leq \|g\|^2$, one has $(u_n, g) \to 0$; an alternative argument: $|(u_n, g)| \leq \|u_n\|\|g_n\| = \|g_n\|$ where $g_n = g_n(x)$ is the restriction of $g(x)$ to the interval $(n-1, n)$, and $\|g_n\| \to 0$ because $\sum_n \|g_n\|^2 = \|g\|^2$

(1.32)
(1) $(-n^2+1)u_n = g_n$, then no solution if the coefficients $g_{\pm 1} \neq 0$. If $g_{\pm 1} = 0$ the solution exists but is not unique (indeed, the coefficients $u_{\pm 1}$ remain arbitrary)
(2) The solution exists and is unique for any $g(t)$

(1.33)
(1) The solution exists and is unique for any $g(t)$
(2) Use the same argument as in Problem 1.25, q.(2)

(1.34)
The answer to this apparent contradiction is simply that in these problems we are using Fourier expansions in spaces as $L^2(0, 2\pi)$ and therefore we are "forced" to look for *periodic* solutions with *fixed* period 2π. Then, if one wants to have the complete solution, one has to add "by hand" the "extraneous" solutions

(1.35)
(1)(*b*) From $\cos^2\varphi = (1+\cos 2\varphi)/2$ one obtains $U(r,\varphi) = (1/2) + (1/2)r^2\cos 2\varphi$
(2) See Problem 1.24, q.(4)

(1.36)
(1) Impose $U = 0$ at $\varphi = 0$
(2) $U(1, 3\pi/2) = -1$, see Problem 1.20, q.(1–2)

(1.37)
(1) Impose $U = 0$ at $\varphi = 0$ and $\varphi = \pi/2$
(2) $U(1, 3\pi/4) = U(1, 7\pi/4) = -1;\ U(1, 5\pi/4) = 1$

(1.38)
(1) $\partial U/\partial r\big|_{r=1} \equiv \sum_{n=1}^{\infty} n(a_n \cos n\varphi + b_n \sin n\varphi) = g_0 + \sum_{n=1}^{\infty} \ldots$, then ...
(2) $g_0 = 0$ is guaranteed by the Gauss law: the integral $\int_0^{2\pi} G(\varphi)\, d\varphi$ is the flux across the circumference of the electric (radial) field which is given just by $\partial U/\partial r$; this flux is zero because $\Delta U = 0$ means that there are no charges at the interior of the circle; the non-uniqueness of the solution corresponds to the property that the potential is defined apart from an additive constant

(1.39)
(1) $\frac{X''}{X} = -\frac{Y''}{Y} = C$, with clear notations, and $X'' = C\,X$ gives $C = -n^2$ thanks to the boundary conditions, etc.
(2) $U(x, y) = \dfrac{\sin x}{e^{-1} - e}\big(\exp(y-1) - \exp(1-y)\big) + \dfrac{\sin 2x}{e^2 - e^{-2}}\big(\exp(2y) - \exp(-2y)\big)$
(3) Only three boundary conditions must be given; the solution is as in (1) but $a_n = 0,\ \forall n$

(1.40)
It is enough to exchange the role of x and y

(1.41)
(2)(*b*) $U(r,\varphi) = (2/5)(r + r^{-1})\cos\varphi$
(*c*) $U(r,\varphi) = (2/3)(-r/2 + 2/r)\cos\varphi + (4/15)(r^2 - r^{-2})\cos 2\varphi$

(1.42)
(1)(*a*) For any $v \in V,\ w \in V^\perp$, one has $(T^+w, v) = (w, Tv) = (w, v') = 0$, with $v' \in V$, which implies $T^+w \in V^\perp$
(*b*) Let $z \in \mathtt{Ker}\,(T^+)$, i.e., $T^+z = 0$, and let x be *any* vector in H, then $0 = (T^+z, x) = (z, Tx)$, which shows that $\mathtt{Ker}\,T^+$ is orthogonal to $\mathtt{Ran}\,T$. The first equality is true also if T is unbounded (recall that T^+ is a closed operator and therefore $\mathtt{Ker}\,(T^+)$ is a closed subspace), the second must be replaced by $\overline{\mathtt{Ker}\,(T)} = \big(\mathtt{Ran}\,(\mathtt{T}^+)\big)^\perp$
(2)(*a*) A counterexample is $T = \begin{pmatrix} 0 & 1 \\ 0 & 0 \end{pmatrix}$
(*b*) (*i*) If $Tv = \lambda v$, then $0 = \|(T - \lambda I)v\|^2 = \big((T - \lambda I)v, (T - \lambda I)v\big)\|^2 =$
$\big((T^+ - \lambda^* I)(T - \lambda I)v, v\big) = \big((T - \lambda I)(T^+ - \lambda^* I)v, v\big) = \|(T^+ - \lambda^* I)v\|^2$,
then $T^+v = \lambda^* v$
(*ii*) From $Tv_1 = \lambda_1 v_1,\ Tv_2 = \lambda_2 v_2$, one gets $(v_2, Tv_1) = \lambda_1(v_2, v_1)$ but also $(v_2, Tv_1) = (T^+v_2, v_1) = \lambda_2(v_2, v_1)$, thanks to (*i*) above, then $(v_2, v_1) = 0$
(3)(*a*) From $Tv = \lambda v$ one has $(v, Tv) = \lambda\|v\|^2$, but (v, Tv) is real: $(v, Tv)^* = (Tv, v) = (v, Tv)$
(3)(*b*) If $\lambda_n \in \mathbf{R}$, expanding x and y as Fourier series of the eigenvectors v_n:
$x = \sum_n a_n v_n,\ y = \sum_n b_n v_n$, one has
$(Tx, y) = (\sum_n \lambda_n a_n v_n, \sum_m b_m v_m) = \sum_n \lambda_n a_n^* b_n = (x, Ty)$
(4)(*a*) Expand any $x \in H$ as Fourier series of the eigenvectors: $x = \sum_n a_n v_n$, then $\|Tx\|^2 = \|\sum_n a_n \lambda_n v_n\|^2 = \sum_n |\lambda_n|^2\,|a_n|^2 \le \sup_n |\lambda_n|^2 \|x\|^2$; then $\|T\| = \sup_n |\lambda_n|$ and possibly $\|T\| = \max_n |\lambda_n|$ when ....
See Problem 1.49 for the first examples, and many other cases in this book.
For the counterexamples, take the matrices

$$T_1 = \begin{pmatrix} 0 & 1 \\ 0 & 0 \end{pmatrix} \quad \text{and} \quad T_2 = \begin{pmatrix} 0 & 1 \\ 0 & 1 \end{pmatrix}$$

(5) The last question: $0 = (x, T^+Tx) = \|T\,x\|^2$ then $Tx = 0$, etc.

(1.43)
(1) From $\|x\| = \|Tx\| = |\lambda|\,\|x\|$ one has $|\lambda| = 1$
If $Tv_n = \lambda_n v_n$ with $|\lambda_n| = 1$ and v_n an orthonormal complete set, then certainly T is invertible; expand now any x and $y \in H$ as Fourier expansion in terms of the eigenvectors: $x = \sum_n a_n v_n,\ \ y = \sum_n b_n v_n$; then $(Tx, Ty) = (\sum_n a_n \lambda_n v_n, \sum_m b_m \lambda_m v_m) = \sum_n a_n^* |\lambda_n|^2 b_n = (x, y)$.
Another proof (using previous problem, q. (2)(*b*)(*i*)): $T^+v_n = \lambda_n^* v_n = \lambda_n^{-1} v_n,\ \forall n$, and this is enough to conclude $T^+ = T^{-1}$
(2) Look for a vector $z \in H$ orthogonal to $U\,e_n$:
$0 = (Ue_n, z) = (Ue_n, UU^{-1}z) = (e_n, U^{-1}z)$ which implies $U^{-1}z = 0$ and $z = 0$

(3) The convergence is ensured by the property $\sum |(e_n, x)|^2 < \infty$; the series converges to the vector Tx
(4) $U^{-1}e_1 = U^+e_1 = \frac{e_1}{\sqrt{2}} + \frac{e_2}{\sqrt{6}} + \ldots + \frac{e_n}{\sqrt{n(n+1)}} + \ldots$, etc., etc. !

(1.44)
(1) Let V_λ be the n-dimensional subspace of the eigenvectors of T with eigenvalue λ. If $v \in V_\lambda$ one has $T(Sv) = STv = \lambda(Sv)$, then $Sv \in V_\lambda$ and $S : V_\lambda \to V_\lambda$. If $n = 1$ then Sv is eigenvector of S. If $n > 1$, the restriction of S to V_λ can be represented by a $n \times n$ matrix, and the known theorems about matrices can be clearly applied: e.g.., there is always *at least* one eigenvector of S; if the restriction to V_λ of S is normal, then .... Nothing can be said, in general, if $n = \infty$.
(2)(*a*) Let $Tv = \lambda v$. Then, e.g.., $TS^2v = (S^2T + 2\sigma S^2)v = (\lambda + 2\sigma)(S^2v)$, etc.; then $S^n v$ (if not zero) is eigenvector of T with eigenvalue $\lambda + n\sigma$
(*b*) $[T^+, S^+] = -\sigma^* S^+$, and if T is Hermitian then S^+v is eigenvector of T with eigenvalue $\lambda - \sigma^*$, etc.; if T is normal and $Tv = \lambda v$, then S^+v is eigenvector of T^+ with eigenvalue $\lambda^* - \sigma^*$, etc.

(1.45)
(1) Imposing idempotency of $P_1 + P_2$ one finds $P_1P_2 + P_2P_1 = 0$, which also gives $P_1P_2 + P_1P_2P_1 = P_1P_2P_1 + P_2P_1 = 0$ and finally the condition $P_1P_2 = P_2P_1 = 0$. This means that H_1 is orthogonal to H_2 and $P_1 + P_2$ projects on the subspace $H_1 \oplus H_2$.

An example: let $\{e_n\,;\, n = 1, 2, \ldots\}$ be an orthonormal complete set in H; let H_1 be generated by $\{e_{4p+1}\,;\, p = 0, 1, 2, \ldots\}$ and H_2 be generated by $\{e_{4q+3}\,;\, q = 0, 1, 2, \ldots\}$; then $H_1 \oplus H_2$ is the subspace generated by e_n with odd indices n.
(2) The condition is $P_1P_2 = P_2P_1$ (not necessarily $= 0$); P_1P_2 projects on $H_{12} = H_1 \cap H_2$, and $H_1 = H' \oplus H_{12}$, $H_2 = H'' \oplus H_{12}$ with $H' \perp H''$.

An example: let $\{e_n\}$ be as in (1), H_1 be generated by $\{e_p\}$ with p multiples of 2, and H_2 by $\{e_q\}$ with q multiples of 3; then P_1P_2 projects on H_{12} = the subspace generated by the e_n where n are the multiples of 6

(1.46)
(1) The subspace on which the projection projects must be finite dimensional
(2) No, no: consider e.g.. the case where T is an unbounded functional
(3) Let w_n be weakly convergent to w; then Bw_n weakly converges to Bw, indeed, $(x, Bw_n - Bw) = (B^+x, w_n - w) \to 0$

CB is compact, indeed Bw_n is weakly convergent, then $C(Bw_n)$ is norm-convergent. BC is compact, indeed Cw_n is norm-convergent, then the same for BCw_n

(1.47)
Norm convergence: $\|T - T_N\| = \sup_{n>N} |\lambda_n| \to 0$
Strong convergence: $\|(T - T_N)x\|^2 = \sum_{n>N} |\lambda_n P_n x|^2 \le M^2 \sum_{n>N} |P_n x|^2 \to 0$

(1.48)
(1) If A is bounded, the conditions are $A^+ = A$ and A positive definite. If A is unbounded, the obvious conditions about its domain must be included
(2) A Cauchy sequence $\{u_n\}$ with respect to the norm induced by the scalar product $< , >$ may be *not* a Cauchy sequence with respect to the norm induced by the scalar product $(,)$: let, e.g.., A be defined by $A\, e_n = e_n/n$ where $\{e_n,\ n = 1, 2, \ldots\}$ is an orthonormal complete set, and let $u_n = e_n$, then $< e_n, e_n >= 1/n$, ...

(1.49)
All the operators are defined on a complete set, then their domain is (at least) dense in H. But (1)*(ii)* cannot be defined on vectors as $x = \sum_n e_n/n \in H$; the same for (2)*(iii)*. The other operators can be extended to the whole space and produce bounded operators, with $\|T\| = \sup_n |c_n|$. In particular, $\|T\| = \max_n |c_n| = 1$ in the cases (1)*(i)*, (2)*(i)*,*(ii)*; $\|T\| = \sup_n |c_n| = 1$ in the cases (1)*(iii)*, (2)*(iv)*. The other operators are unbounded. In the cases (1)*(iii)* and (2)*(iv)* there is no x_0 such that $\|Tx_0\| = \|T\|\,\|x_0\|$. The eigenvalues of (2)*(i)* have degeneracy 14, those of (2)*(iv)* are doubly degenerate (if $n \neq 0$). The range of (1)*(i)* contains a complete set, then is dense in H, but it is not the whole space: indeed, e.g.., the vector $y = \sum_n e_n/n \in H$ cannot be the image of any vector in H (the inverse T^{-1} of this operator is just the operator (1)*(ii)*). The ranges of the operators (1)*(ii)*, *(iii)*, (2)*(i)*, *(ii)* are the whole Hilbert space H: indeed, e.g.., for the case (1)*(ii)*, taken any $y \in H,\ y = \sum_n b_n e_n$, there is $x = \sum_n (b_n/n) e_n \in H$ such that $Tx = y$; finally, the ranges of the operators (2)*(iii)*, *(iv)* are proper subspaces of H, indeed for $n = 0$ and resp. $n = \pm 1$ ...

(1.50)
(1) T_N projects on a $2N + 1$-dimensional subspace; it is compact
(2) $\|(I - T_N)x\|^2 = \sum_{|n|>N} |x_n|^2 \to 0$, but $\|I - T_N\| = \sup_x \frac{\|(I-T_N)x\|}{\|x\|} = 1,\ \forall N$ (choose, e.g., $x = e_{N+1}$), then: strong, no norm, convergence

(1.51)
(1) S_N has the zero eigenvalue infinitely degenerate, the eigenvalues ± 1 with eigenvectors $e_j \pm e_{-j}$, $(1 \le j \le N)$ and the eigenvalue 1 with eigenvector e_0
(2) See previous problem
(3) $S_\infty f(x) = f(-x)$

(1.52)
(1) $(v_n, v_m) = \delta_{nm}$, but $\{v_n\}$ is not complete: e.g.. $e_1 - e_2$ is orthogonal to all the $\{v_n\}$
(2)*(a)* $\mathtt{Ran}\, T$ is the Hilbert space of the vectors $v = \sum a_n v_n$ with $\{a_n\} \in \ell^2$; the orthogonal complement is given by the vectors $z = \sum c_n w_n$ with $\{c_n\} \in \ell^2$ and $w_n = (e_{2n-1} - e_{2n})/\sqrt{2}$
(b) $(Tx, Ty) = \left(\sum_n x_n v_n, \sum_m y_m v_m\right) = (x, y)$, with clear notations, then T is isometric but not unitary (it is not surjective)
(c) Let $x = \sum_n a_n e_n$; the equations for the first components of $Tx = \sum_n a_n v_n = \lambda x$ are $a_1 = \lambda\sqrt{2}\, a_1,\ a_1 = \lambda\sqrt{2}\, a_2,\ a_2 = \lambda\sqrt{2}\, a_3, \ldots$ and it is easy to conclude, using also the property of the eigenvalues of isometric operators (see Problem 1.44,q.(1)),

that there are no solutions apart from $a_n = 0,\ \forall n$. Then there are no eigenvectors and $\mathtt{Ker}\,T = \{0\}$

(1.53)
(1) $Tx = T(a_1, a_2, \ldots) = (0, a_1, a_2, \ldots);\ \ Sx = (a_2, a_3, \ldots)$
(2) T injective, not surjective; the opposite for S
(3) $\|T\| = 1$, clearly. $\|Sx\|^2 = \sum_{n=2}^{\infty} |a_n|^2 \le \|x\|^2$, then $\|S\| \le 1$, but choosing, e.g.., $x = e_2$, one has $\|Se_2\| =$..., then $\|S\| = 1$
(4) Let $x = \sum_n a_n e_n,\ \ y = \sum_n b_n e_n$, then $(y, T^+x) = (Ty, x) = \sum_{n=1}^{\infty} b_n^* a_{n+1} = (y, Sx)$;
(5) $(Tx, Ty) = \sum_{n=1}^{\infty} b_n^* a_n = (x, y)$
$T^+T = I;\ TT^+$ projects on the subspace of vectors $x \in H$ with $(e_1, x) = 0$
(6) The eigenvalue equation $Tx = \lambda x$ becomes $(0, a_1, a_2, \ldots) = \lambda(a_1, a_2, \ldots)$ then the first equations are $0 = \lambda a_1,\ a_1 = \lambda a_2$, etc. and it is easy to conclude that the only solution is the trivial one $x = 0$ (recall the property of isometric operators, see q. (2)(*c*) of the previous problem). $Sx = \lambda x$ is $(a_2, a_3, \ldots) = \lambda(a_1, a_2, \ldots)$, so $a_2 = \lambda a_1,\ a_3 = \lambda a_2, \ldots, a_{n+1} = \lambda^{n-1} a_1, \ldots$, and this produces a vector $\in H$ if and only if $|\lambda| < 1$
(7) No, indeed (for instance) T transforms the weakly convergent sequence $\{e_n\}$ into itself; a similar result for S

(1.54)
(1) T (and T^+, of course) is unitary. T and S do not possess eigenvectors (see previous problem)
(2) The eigenvalue equation $(\exp(ix) - \lambda) f(x) = 0$ has only the trivial solution $f(x) = 0$ (for any λ, the equation $\exp(ix) = \lambda$ admits, at most, a finite number of solutions, then, ..., see Problem 1.14, q.(2))

(1.55)
(1) Similar answers to those of Problem 1.53. For the eigenvectors of the form $g(x)$, see Problem 1.53, q.(6); the functions $h(x) = \exp(-\alpha x)$ are eigenvectors for any $\alpha > 0$ with $\lambda = \exp(-\alpha)$
(2) T and S are unitary operators. $Tf(x) = f(x-1) = \lambda f(x)$ implies (recall that $|\lambda| = 1$, see previous problems) $|f(x-1)| = |f(x)|$, i.e., $|f(x)|$ periodic, but no function in $L^2(\mathbf{R})$ can satisfy this condition

(1.56)
$Tx = T(\sum_{n=1}^{\infty} a_n e_n) = T(a_1, a_2, \ldots) = (0, a_1, 0, a_2, 0, \ldots)$.
$T^+x = Sx = (a_2, a_4, \ldots)$.
TT^+ projects on the subspace of vectors $x \in H$ with $(e_{2n+1}, x) = 0,\ \forall n = 0, 1, 2, \ldots$
$Tx = \lambda x$ becomes $(0, a_1, 0, a_2, 0, \ldots) = \lambda(a_1, a_2, \ldots)$ and there are no eigenvectors .
$T^+x = Sx = \lambda x$ becomes $(a_2, a_4, \ldots) = \lambda(a_1, a_2, \ldots)$ and here one obtains

$a_2 = \lambda a_1,\ a_4 = \lambda a_2,\ a_6 = \lambda a_3$ etc. Then for any $|\lambda| < 1$, an eigenvector is $a(1, \lambda, 0, \lambda^2, 0, \ldots)$, another is $a(0, 0, \lambda, 0, 0, \lambda^2, 0, \ldots)$ and so on. For each λ such that $|\lambda| < 1$ there are infinitely many eigenvectors!

(1.57)
T has no eigenvectors.

$\|T\| = \|S\| = 1/\sqrt{2}$. The functions x^α are eigenvectors of S with eigenvalue $\lambda_\alpha = 1/2^{\alpha+1}$; recall that $x^\alpha \in L^2(0, 1)$ only if $\alpha > -1/2$, then $\sup_\alpha \lambda_\alpha = 1/\sqrt{2}$.

T is not compact because (for instance) it transforms the weakly convergent sequence $\{\sin n\pi x\}$ into another weakly (not strongly) convergent sequence. A similar result for S

(1.58)
(1) T^N converges weakly to 0:

$$|(g, T^N f)| = \left| \int_N^\infty g^*(x) f(x - N) dx \right| \le \|f\| \|g^{(N)}\| \to 0$$

where $g^{(N)} = g^{(N)}(x)$ is the "queue" of the function $g(x)$, i.e. $g^{(N)}(x) = g(x)$ for $x > N$ and $g^{(N)}(x) = 0$ for $0 < x < N$.

On the other hand $\|T^N f\| = \|f\|$, then T^N does not converge strongly.

For any f one has

$$\|S^N f\|^2 = \int_0^\infty |f(x + N)|^2 dx = \int_N^\infty |f(x)|^2 dx \to 0$$

being $f(x) \in L^2$; then $S^N \to 0$ strongly, not in norm because $\|S^N\| = 1$
(2) T^N converges weakly to 0:

$$(g, T^N f) = \int_{-\infty}^{+\infty} g^*(x) f(x - N)\, dx = (2\pi)^{-1} \int_{-\infty}^{+\infty} \widehat{g}^*(y) \widehat{f}(y) \exp(iNy)\, dy \to 0$$

where $\widehat{f}(y),\ \widehat{g}(y)$ are the Fourier transforms of $f(x),\ g(x)$, thanks to the Riemann-Lebesgue theorem, because $\widehat{g}^*(y)\widehat{f}(y) \in L^1(\mathbf{R})$. No strong convergence, indeed $\|T^N f\| = \|f\|$. Obviously the same results hold for S^N
(3) $T^N \to 0$ weakly: given $x = \sum a_n e_n,\ y = \sum b_n e_n$, one has

$$|(y, T^N x)| = \left| \sum_{n>N} b^*_{n+N} a_n \right| \le \|x\| \|y^{(N)}\| \to 0$$

where $y^{(N)}$ is the "queue" of the vector y, i.e. $y^{(N)} = \sum_{n>N} b_n e_n$; no strong convergence.

$S^N \to 0$ strongly, indeed, for any $x = \sum_{n=1}^\infty a_n e_n$ one has

$$\|S^N x\|^2 = \sum_{n>N} |a_n|^2 \to 0$$

being $\{a_n\} \in \ell^2$;
(4) $T^N \to 0$ weakly:

$$(g, T^N f) = \int_0^{2\pi} g^*(x) f(x) \exp(iNx) dx \to 0$$

indeed this can be viewed as the Nth Fourier coefficient of the function $g^*(x)f(x) \in L^1(0, 2\pi)$. A Riemann-Lebesgue-type theorem ensures that these coefficients vanish as $N \to \infty$. Clearly, no strong convergence. The same for S^N
(5) A similar argument and the same result as in q. (4): given $a \equiv (\ldots, a_{-1}, a_0, a_1, a_2, \ldots)$ and $b \equiv (\ldots, b_{-1}, b_0, b_1, b_2, \ldots)$ and introducing the functions $f = \sum_{n\in\mathbf{Z}} a_n e_n$, $g = \sum_{n\in\mathbf{Z}} b_n e_n$, where $e_n = \exp(inx)(2\pi)^{-1/2}$, the scalar product $(b, T^N a) = \sum_{n\in\mathbf{Z}} b^*_{n+N} a_n$ can be thought as the scalar product of two functions $\in L^2(0, 2\pi)$, etc.

(1.59)
(1) *(a)* $Tx = T\big(\sum a_n e_n\big) = \sum (a_n - a_{n+1}) e_n$, then $Tx = 0$ if $a_{n+1} = a_n$, which implies $a_n = 0$, $\forall n$, and $\mathtt{Ker}\, T = \{0\}$.
(b) $\|Tx\| = \|\sum a_n e_n - \sum a_n e_{n-1}\| \le 2\|x\|$
(c) $\|T(e_0 + \ldots + e_k)\| = \|e_0 - e_{k+1}\| = \sqrt{2}$ whereas $\|e_0 + \ldots + e_k)\| = \sqrt{k+1}$, then T^{-1} is unbounded
(2) *(a)* Yes. Yes: see Problem 1.13
(b) $T = \alpha I + \beta S$, where I is the identity and S is a "shift" operator: no eigenvectors, see Problem 1.54
(3) *(a)* $\|T\| = \sup_x |\varphi(x)| = \sup_x |\alpha - \beta \exp(-ix)| = |\alpha| + |\beta|$;
(c) the condition is $|\alpha/\beta| \ne 1$ and $\|T^{-1}\| = \big||\alpha| - |\beta|\big|^{-1}$
(d) $g_1(x) = 1 \notin \mathtt{Ran}\, T$ because $Tf_1 = g_1$ would imply $f_1 = 1/\big(1 - \exp(-ix)\big)$ which $\notin L^2(0, 2\pi)$ $\big($this does not contradict the density of $\mathtt{Ran}\, T$ in L^2, indeed the function $g_1(x) = 1$ can be approximated, in the sense of the L^2 norm, by functions which tend suitably to zero at $x = 0$ and $x = 2\pi$: see Problem 1.6$\big)$. Instead $g_2(x) = \sin x \in \mathtt{Ran}\, T$ because $\sin x/\big(1 - \exp(-ix)\big) \in L^2(0, 2\pi)$

(1.60)
(1) Put $c = \sum c_n e_n$, then $Tx = x_0(c, x) + x$ and $\|Tx\| \le \big(\|x_0\|\,\|c\| + 1\big)\|x\|$
(2) The eigenvalues equation is $x_0(c, x) = (\lambda - 1)x$; then: if $\lambda = 1$ the eigenvectors are the vectors orthogonal to c (and they generate a subspace H_1 with codimension 1); if $\lambda \ne 1$, then x must be proportional to x_0 $\big($with eigenvalue $\lambda = 1 + (c, x_0)\big)$. Now, if $x_0 \notin H_1$, the eigenvectors are a complete set for H, otherwise, i.e., if $(c, x_0) = 0$, the eigenvectors generate only H_1 and are not a complete set.
(3) From (1) one has $\|T\| \le 1 + \|x_0\|^2$, choose now $x = x_0$ to conclude $\|T\| = 1 + \|x_0\|^2$. An alternative argument: in the case $c = x_0$, the eigenvectors are a complete orthogonal set, then $\|T\| = \sup|\text{eigenvalues}| = \max\{1,\ 1 + \|x_0\|^2\}$

(1.61)
(1) *(a)* $|\alpha_n| = 1$; *(b)* $\sup_n |\alpha_n| < \infty$; *(c)* $T^2 e_n \propto e_{n+2}$ then for no α_n the operator T can be a projection (apart from the case $T = 0$!). Or: $T \neq T^+$, see (2)
(2) $T^+ e_n = \alpha^*_{n-1} e_{n-1}$
(3) *(a)* $\|T\| = 2$
(b) Ker T is generated by e_n with $n = 4k - 1$; e.g., e_3 belongs to Ker T, but also to Ran T, indeed $T\, e_2 \propto e_3$

(1.62)
(1) If $x = \sum_{n=1}^N a_n e_n$ one has $Tx = \big(\sum_{n=1}^N a_n\big) x_0$ then T is unbounded and there are vectors $\notin$ domain of T (which is however dense in H !)
(2) Ker $T = \ell^{(0)}$, dense in H ! (see Problem 1.16)
(3) e.g.. $z_n = e_1 - (e_2 + \dots e_{n+1})/n \in$ Ker T and $w_n = e_1 + e_n/n$
(4) No, as shown, e.g.., by q. (3)

(1.63)
(1)-(3)-(4) Put $T' = T - \alpha I$ and compare with the previous problem
(2) $\alpha = -\beta$, Ker T is the 1-dimensional subspace generated by e_1

(1.64)
(1) *(a)* $\sup_n |c_n| < \infty$, *(b)* $|c_n|^2 = |c_{-n}|^2$, *(c)* $c_n = c^*_{-n}$, *(d)* $c_n c_{-n} = 1$
(2) The 2–dimensional subspaces are generated by $e_{\pm n}$. For any $n \neq 0$ one finds two independent eigenvectors, then the eigenvectors form a complete set in H, not necessarily orthogonal
(3) *(a)* eigenvectors orthogonal, eigenvalues not necessarily real;
(b) eigenvalues $\pm 1/\sqrt{n}$; no surprise, indeed T is not normal

(1.65)
(1) The problem is 1- or 2-dimensional; see the previous problem
(2) $c \neq \pm n^2$
(3) Writing $T + \alpha I$, the required norms are given by $\sup_n \left|\frac{1}{\alpha \pm n^2}\right| = \frac{1}{\inf_n |\alpha \pm n^2|}$, then resp. $1/4$, 1, $1/\sqrt{2}$

(1.66)
(1) e_0 with zero eigenvalue; $e_n \pm e_{-n}$ with eigenvalues $\pm 1/n^2$, not degenerate
(2) Ran T is orthogonal to the subspace of constants, but is dense in (not coinciding with) the complementary subspace, which is just the closure $\overline{\text{Ran}\, T}$; e.g.., $g(x) = \sum_{n \neq 0} (1/n) \exp(inx) \notin$ Ran T, see also q. (3)*(c)* below)
(3) $T e_0 = 0$ implies that the equation does not admit solution if $g_0 = (e_0, g) \neq 0$; the presence of the coefficients $1/n^2$ implies that $g(x)$ must be at least two times differentiable with $g''(x) \in L^2$. Then: *(a)* (not unique) solution $f(x) = 2\cos(2x) +$ const.
(b) $\cos^4(x) \geq 0$, then $g_0 \neq 0$, no solution; *(c)* $g_0 = 0$ but $g(x)$ is *not* two times differentiable, then no solution

(4) T^N converges to the operator S_1 defined by $S_1\, e_{\pm 1} = e_{\mp 1}$ and $S_1 = 0$ on the other e_n. Indeed $\|T^N - S_1\| = \max_{n\neq 0} |\text{eigenvalues}| = 1/2^{2N} \to 0$

(1.67)
(1) One has $\|z\| = 1/\sqrt{3}$ and $Tx = (e_1+e_2)(z,x)$, then: (*a*) $\|T\| = \sqrt{2/3}$
(*b*) (e_1+e_2) with eigenvalue $3/4$ and any vector orthogonal to z with zero eigenvalue. (Notice that $\sqrt{2/3} > 3/4$)
(*c*) $T^+x = z\big((e_1,x)+(e_2,x)\big)$
(2) If $\alpha_n \in \ell^2$, $\beta_n \in \ell^2$, put $\widehat{\alpha} = \sum_n \alpha_n^* e_n$ and $\widehat{\beta} = \sum_n \beta_n^* e_n$: then $Tx = (\widehat{\alpha},x)e_1 + (\widehat{\beta},x)e_2$ is bounded (and conversely)
(3)(*a*) If $\widehat{\alpha}$ and $\widehat{\beta}$ are linearly dependent, e.g., $\widehat{\alpha} = c\widehat{\beta}$, then the range of T is 1-dimensional and is given by the multiples of $e_1 + ce_2$.
(*b*) If $\mathtt{Ran}\, T$ is 1-dimensional, $\mathtt{Ker}\, T$ is given by the vectors orthogonal to $\widehat{\alpha}$, then $\widehat{\alpha}$ must be proportional to $e_1 + c\, e_2$; if $\widehat{\alpha}$ and $\widehat{\beta}$ are linearly independent, $\mathtt{Ran}\, T$ is generated by e_1, e_2, $\mathtt{Ker}\, T$ is orthogonal to $\mathtt{Ran}\, T$ if both $\widehat{\alpha}$ and $\widehat{\beta}$ are combinations of e_1 and e_2.
(4) One has $\|T_N x\| = \|e_1+e_2\||(z_N,x)| = \sqrt{2}|(z_N,x_N)| \to 0$ where x_N is the "queue" of the vector x, i.e., $x_N = \sum_{n>N} a_n e_n$. But $\|T_N\| = \sqrt{2}\|z_N\| = \sqrt{2/3}$, then strong convergence to zero. Instead, only weak convergence to zero for T_N^+: indeed $\|T_N^+ x\| = \|z_N\||a_1+a_2|$, on the other hand $(y, T_N^+ x) = (a_1+a_2)(y,z_N) \to 0$; alternatively: $(y, T_N^+x) = (T_N y, x)\ldots$

(1.68)
(1) The vectors $\{e_n,\ |n| \le N\} \cup \{(e_n \pm e_{-n}),\ |n| > N\}$ are a complete orthogonal set of simultaneous eigenvectors of S and T_N.
(2) S and $T_N S$ are compact: indeed they admit a complete orthogonal set of eigenvectors with eigenvalues $\to 0$.
(3) $T_N \to I$ strongly: $\|(I-T_N)x\|^2 = \sum_{|n|>N} |x_n|^2 \to 0$, but $\|I - T_N\| = 1$, with clear notation. $T_N S \to S$ in norm: $\|S - T_N S\| = \frac{2(N+1)^2}{1+(N+1)^4} \to 0$

(1.69)
(1) Domain = $L^2(-a,a)$; $\mathtt{Ker}\, T$ = {the functions orthogonal to the constants = the functions with zero mean value}, with dimension ∞ and codimension 1.
$\mathtt{Ran}\, T$ is 1-dimensional.
(2) $\lambda = 0$ with eigenvectors the vectors $\in \mathtt{Ker}\, T$; the other eigenvector is $h(x)$ with eigenvalue $\lambda = \int_{-a}^{+a} h(x)dx$, not degenerate $\big($unless $\int_{-a}^{+a} h(x)dx = 0\big)$.
(3) $\|Tf\| = \|h\|\,|(1,f)| \le \|h\|\,\|f\|\sqrt{2a}$, choosing $f(x) = 1$ one concludes $\|T\| = \sqrt{2a}\|h\|$
(4) $T^+ f = (h,f)\chi_a(x)$ where $\chi_a(x)$ is the characteristic function of the interval $(-a,a)$. Eigenvalue $\lambda = 0$ with eigenvectors the functions $g(x)$ such that $(h,g) = 0$, and eigenvector $\chi_a(x)$ with eigenvalue $\lambda = \int_{-a}^{+a} h(x)dx$ $\big($see (2)$\big)$
(5) $T^2 f = h(1,h)(1,f)$

(1.70)
See previous problem. If $h(x) \notin L^2$, the operator $T : L^2 \to L^2$ cannot be defined correctly; if $h(x) \notin L^1$, the eigenvalue $\lambda \neq 0$ does not exist
The domain of T is $L^1 \cap L^2$. $\mathtt{Ker}\, T$ is dense in $L^2(\mathbf{R})$. See Problems 1.5 and 1.62

(1.71)
(1) See Problem 1.69
(2)*(a)* $T^+x = (v, x)w$. See Problem 1.69
(b) $v = w$ and $\|v\| = 1$ (T projects on the 1-dimensional space generated by v)
(c) $(v, w) = 0$

(1.72)
(1) T_N is the projection of $f(x)$ on the first $|n| \le N$ components of the Fourier expansion on the set $\{\exp(inx)/\sqrt{2\pi}\}$
Strong convergence to the identity operator, indeed $\|(I - T_N)f\|^2 = \sum_{|n|>N} |f_n|^2 \to 0$, with clear notations, but $\|I - T_N\| = \sup_f \frac{\sum_{|n|>N} |f_n|^2}{\|f\|^2} = 1$ (choose, e.g., $f(x) = e_{N+1}$)
(2) *(a)* The identity ! *(b)* The derivative; *(c)* $g(x) \in C^\infty$ $\big($see Problem 1.24, q.(4)$\big)$

(1.73)
(1) Only $A_n^{(-)}$ and $C_n^{(-)}$ are projections:
$A_n^{(-)} f = e_n(e_n, f)$ projects on the 1-dimensional space generated by $e_n = \exp(inx)/\sqrt{2\pi}$
$C_n^{(-)} f = w_n(w_n, f) + v_n(v_n, f)$ projects on the 2-dimensional space generated by $w_n = \cos nx/\sqrt{\pi}$, $v_n = \sin nx/\sqrt{\pi}$
(2) E.g.: $A_n^{(+)} f = \exp(inx)\big(\exp(-iny), f(y)\big) = e_n(e_{-n}, f)$, then see Problem 1.72;
$B_n^{(+)}$ has eigenvectors $v_n \pm w_n$, and all v_m, w_m with $m \neq n$
(3) Strong convergence to 0. E.g.,
$\|B_n^{(+)} f\|^2 = |(w_n, f)|^2 + |(v_n, f)|^2 \to 0$, being coefficients of a Fourier expansion, but $\|B_n^{(+)}\| = 1$ (choose, e.g.., $f = v_n$)

(1.74)
(1–2) T_∞ is the projection on the subspace of odd functions. For the convergence properties, see, e.g.., Problems 1.72 or 1.73
(3) T_N are compact, T_∞ is not (indeed, the weakly convergent sequence $\{\sin nx\}$ is mapped by T_∞ into itself)
(4) T_∞ is the identity, the same as in Problem 1.50

(1.75)
The functions $1, x, \ldots, x^N$ are not orthogonal
(1) $\mathtt{Ran}\, T$ has dimension $N + 1$, $\mathtt{Ker}\, T$ is ∞-dimensional
(2) Only $f_1 = x$ is eigenvector
(3) No: e.g.., $T_2^2(1) \neq T_2(1)$
(4) Besides the eigenvalue zero, there is the eigenvalue $2/3$ with eigenvector x, and the eigenvalues $(18 \pm 2\sqrt{61})/15$ with eigenvectors $5 + (-6 \pm \sqrt{61})x^2$

(1.76)
(1) $T_N/2N$ projects on the 1-dimensional space generated by $\chi_N(x)$
(2) Let $N > M$, then $T_N T_M f = 2M\chi_N(\chi_M, f)$ and $T_M T_N f = 2M\chi_M(\chi_N, f)$
(3) The eigenvalues are 0 and 4 for both operators, the eigenvectors are different ...
(4) Yes, clearly! A simple example is $f(x) = \cos \pi x$ for $|x| < k$ and $= 0$ elsewhere, where k is any integer

(1.77)
(1) $|c_n| = |(\chi_n, f)| = |(\chi_n, f_n)| \le \|f_n\|$, where $f_n = f_n(x)$ is the restriction of $f(x)$ to the interval $(n-1, n)$, then $\sum_n \|f_n\|^2 = \|f\|^2$, see Problem 1.31
(2) The N-dimensional subspace generated by $\chi_1, \dots, \chi_N$ with eigenvalue 1 and the orthogonal space with zero eigenvalue; T_N is a projection for each N
(3) *(a)(b)* T_∞ is defined in the whole space, indeed $T_\infty f = \sum_{n=1}^{\infty} \chi_n c_n$ and $\|T_\infty f\|^2 = \sum |c_n|^2 < \infty$; on the other hand, $\|T_\infty f\|^2 = \sum |c_n|^2 = \sum |(\chi_n, f)|^2 \le \sum \|f_n\|^2 = \|f\|^2$, etc., then $\|T_\infty\| = 1$
(c) Not compact because the eigenvalue 1 is infinitely degenerate (all the χ_n are eigenfunctions with eigenvalue 1; alternatively, because the weakly convergent sequence χ_n is mapped into itself by T_∞
(4) Strong convergence: $\|(T_\infty - T_N)f\|^2 = \sum_{n>N} |c_n|^2 \to 0$, but $\|T_\infty - T_N\| = 1$
(5)*(a)* $\|S_\infty - S_N\| = 1/(N+1) \to 0$
(b) S_∞ compact because its eigenvalues $1/n$ are non-degenerate and $\to 0$

(1.78)
With $e_n = \sin nx/\sqrt{\pi}$, the operator is $T f = \sum (1/n)(e_n, f)e_n$
(1) $T(\exp(2ix)) = (i/2)\sin 2x$
(2) e_n are eigenfunctions with eigenvalue $1/n$, and all even functions with zero eigenvalue. T is then compact.
(3) $T = \sum \frac{1}{n} P_n$ where P_n projects on the 1-dimensional space generated by e_n.
(4) If $g(x) \in \mathtt{Ran}\ T$ then $g(x)$ is a continuous function. Indeed, recall that a series of the form $\sum_n \alpha_n \sin nx$ is certainly uniformly convergent (and then converges to a continuous function) if $\sum_n |\alpha_n| < \infty$, see Problem 1.24. This condition is satisfied in this case, see the argument used in Problem 1.25. Conversely, e.g.. $g(x) = \sum_n n^{-4/3} \sin nx$ is continuous but $\notin \mathtt{Ran}\ T$. The closure $\overline{\mathtt{Ran}\ T}$ coincides with the Hilbert subspace of the odd functions

(1.79)
(1) $\|T\| = 1$ and $f_0(x) = \sin x$
(2) $\lambda = 0$ is the only eigenvalue
(3) $g(x) \in C^0$ (see previous problem); dg/dx has Fourier coefficients $b_n \propto (\sin nx, f)$ then $dg/dx \in L^2$ but in general not continuous; $(d/dx)T f$ has Fourier coefficients $\propto b_n$ then the operator is bounded
(4) $\mathtt{Ran}\ T$ contains only even and zero mean-valued functions, then the solution exists if $\beta = 0$, $\alpha = -\pi/2$

(1.80)
(1) $T_n f = (\pi/2) e_n (e_n - e_0, f)$; $\|T_n\| = \pi/\sqrt{2}$, indeed, $\|T_n f\| = (\pi/2)|(e_n - e_0, f)| \le (\pi/2)\sqrt{2}\, \|f\|$ and choosing $f(x) = e_n - e_0 \dots$
(2) Ran T is 1-dimensional, ..., Ker T and Ran T are not orthogonal.
(3) The eigenvalues are 0 and $\pi/2$
(4) No: $T_n f = 0$ implies $(e_0, f) = (e_n, f)$, $\forall n$, then only $f = 0$
(5) $T_n \to 0$ only in weak sense

(1.81)
(1) $\|T\| = a$; $f_\varepsilon(x) = \begin{cases} 1 \text{ for } a - \varepsilon < x < a \\ 0 \text{ elsewhere} \end{cases}$
(2) $(x - \lambda) f(x) = 0$ implies that $f(x) = 0$ apart from a single point at most, which means $f(x) = 0$ as well-known
(3) The inverse of $T f = (x - \sigma) f = g$ is $T^{-1} g = g(x)/(x - \sigma)$, which is unbounded if $0 \le \sigma \le a$; the spectrum is then the closed interval $[0, a]$

(1.82)
(1) E.g. $f(x) = x/(1 + x^2) \notin D(T)$. $D(T)$ is however dense in H, indeed $D(T)$ certainly contains functions which behave as $1/|x|^\alpha$ for any $\alpha > 3/2$ when $|x| \to \infty$; but it is known that the set of functions rapidly going to zero at the infinity is dense in H. This implies that also operators with $\varphi = x^a$ for any $a > 0$ possess dense domain
(2) $f_n(x) = \begin{cases} 1 \text{ for } n < x < n + 1 \\ 0 \text{ elsewhere} \end{cases}$
(3) Any (continuous) function $f(x) \ne 0$ in a neighborhood of $x = 0$ does not belong to Ran T; Ran T is however dense in H (the closure $\overline{\text{Ran}\, T} = H$). This can be shown in several ways: Ran T contains functions which behave as $|x|^\alpha$ for any $\alpha > 1/2$ when $x \to 0$, and these functions can certainly approximate in norm L^2 any function belonging to L^2 (see also Problem 1.6). Recall also that if $\{e_n(x)\}$ is an orthonormal complete system in H, then $u_n(x) = x\, e_n(x)$ is a complete system in H, and $u_n(x) \in$ Ran T, therefore.... As another useful argument, notice that any function $g(x) \in L^2$ can be approximated (in the norm L^2, of course) by a "truncated" function

$$g_\varepsilon(x) = \begin{cases} 0 & \text{for } |x| < \varepsilon \\ g(x) & \text{for } |x| > \varepsilon \end{cases}$$

for ε "small" enough, and observing that functions as $g_\varepsilon(x)$ clearly belong to Ran T.
(4) All the above arguments also hold for operators with $\varphi(x) = x^a$, $\forall a > 0$

(1.83)
(1) $f_n(x) = \begin{cases} 1 \text{ for } 1/n < x < 1 \\ 0 \text{ elsewhere} \end{cases}$
(2) The role of domain and that of range of this operator are exchanged with respect to the ones of the operator in the previous problem

(1.84)
(*a*) $\sup_{x\in\mathbf{R}} |\varphi(x)| = M < \infty$; $\|T\| = M$, indeed

$$\|T f\|^2 = \int_{-\infty}^{+\infty} |\varphi(x) f(x)|^2 dx \le \sup_{x\in\mathbf{R}} |\varphi(x)|^2 \|f\|^2 \quad \text{etc.}$$

see Problem 1.81 q.(1). Assume that there is some point x_0 such that $|\varphi(x_0)| = M$ and there is some interval J containing x_0 where $|\varphi(x)| = M$: in this case, $f_0(x)$ is any function with support $\subset J$
(*b*) $\varphi(x)$ real; for any $\varphi(x)$; $|\varphi(x)| = 1$;
(*c*) $\varphi(x) = 1$ for x in any subset J (even if not connected) $\subset \mathbf{R}$, and $\varphi(x) = 0$ in $\mathbf{R}\setminus J$
(*d*) $\sup_{x\in\mathbf{R}} |\varphi(x)| = +\infty$
(*e*) if $\inf_{x\in\mathbf{R}} |\varphi(x)| = m > 0$, then $\|T^{-1}\| = 1/m$
(*f*) $\varphi(x)$ possesses isolated zeroes and/or $\inf_{x\in\mathbf{R}} |\varphi(x)| = 0$
(*g*) $\varphi(x)$ must be constant in one or more intervals; the eigenfunctions are all the functions with support in these intervals
(*h*) for no $\varphi(x)$ the operator T is compact: choose any interval J where $\inf_{x\in J} |\varphi(x)| = m > 0$ and any sequence $w_n(x) \in L^2(J)$ weakly (but not norm) convergent to 0; then $T w_n$ is not norm convergent, indeed $\|\varphi\, w_n\| \ge \inf_{x\in J} |\varphi(x)|\, \|w_n\| = m\, \|w_n\|$

(1.85)
See previous problem

(1.86)
(1–3) See Problem 1.84
(3) $H = \overline{\mathtt{Ran}\, T}$, i.e., the closure of $\mathtt{Ran}\, T$
(4) $T^N \to 0$ in strong sense, thanks to Lebesgue theorem: $x^{2N}/(1+x^2)^N \to 0$ pointwise, and $|f(x)|^2 \in L^1$; but $\|T^N\| = 1$

(1.87)
T admits the eigenvalues 0 and 1 infinitely degenerate.

E.g. $1/(1+x)$ for $x \ge 0$ and $= 0$ for $x < 0 \notin \mathtt{Ran}\ T$, but $\mathtt{Ran}\ T$ dense in $L^2(0, +\infty)$, see Problem 1.82. $\mathtt{Ran}\, T$ is orthogonal to $\mathtt{Ker}$, but it is not a Hilbert space; one has $L^2(\mathbf{R}) = \mathtt{Ker}\, T \oplus \overline{\mathtt{Ran}\, T}$.

$T^N \to P$, where P is the projection on the subspace $L^2(0, 1)$, in strong sense, see Problems 1.84 and 1.86

(1.88)
(1) $|\rho| \ne 1$
(2) (*a*) $\|(T + 2i\, I)^{-1}\| = \dfrac{1}{\inf_{x\in(0,2\pi)} |\exp(ix) + 2i|} = 1$, the same in (4)(*a*)
(3) (*a*) $\|(T + 2i\, I)^{-1}\| = 1/\sqrt{5}$
In all cases T^N converges weakly to 0, see Problem 1.58, q.(4)

(1.89)
(1) $T^+ f = \big(1 - \alpha^* \exp(-ix)\big) f(x)$; T is normal
(2) $\|T\| = 1 + |\alpha|$; $|\alpha| \neq 1$
(3) If $|\alpha| < 1$, the series converges in $L^2(0, 2\pi)$ and uniformly to $1/\big(1 - \alpha \exp(ix)\big)$
(4)

$$I - A_N = 1 - \big(1 - \alpha \exp(ix)\big) \sum_{n=0}^{N} \alpha^n \exp(inx) = 1 - \sum_{n=0}^{N} \alpha^n \exp(inx) +$$

$$\sum_{n=0}^{N} \alpha^{n+1} \exp(i(n+1)x) = \alpha^{N+1} \exp(i(N+1)x)$$

then $\|I - A_N\| = |\alpha^{N+1}| \to 0$ and $\sum_{n=0}^{N} S^n \to (I - S)^{-1}$

(1.90)
(1) $T^+ = i\, d/dx$ in the domain D' of the functions which are bounded at $x = 0$ and $x = 1$. Indeed,

$$(g, Tf) = (g, i f') = i \int_0^1 g^* f' \, dx = \int_0^1 (ig')^* f \, dx + i\big[g^*(1) f(1) - g^*(0) f(0)\big] = (Tg, f)$$

and the assumptions on D_0 and D' give $g^*(1) f(1) = g^*(0) f(0) = 0$
(2) In D_α, the term in [...] disappears if $g^*(1) f(1) = |\alpha|^2 g^*(0) f(0)$, i.e., $|\alpha| = 1$, and $T = T^+$ in the common domain D_α
(3) Look for eigenfunctions of the form $\exp(-i\lambda x)$ with eigenvalue λ: imposing the boundary condition $\exp(-i\lambda) = \alpha = \exp(a + i\gamma)$ one finds the eigenfunctions $u_n(x) = \exp(x\, a + i\, x\, \gamma + 2n\pi i\, x)$, $n \in \mathbf{Z}$, with eigenvalue $\lambda = i\, a - \gamma + 2n\pi$. The eigenfunctions are in general not orthogonal, but a complete system, for all α.
(4) If $|\alpha| = 1$ then $a = 0$, $\lambda = \gamma + 2n\pi$ and $u_n(x)$ are orthogonal
(5) All domains are clearly dense in $L^2(0, 1)$

(1.91)
(1) The eigenfunctions are $u_k(x) = \exp(i\, kx)$, $k \in \mathbf{Z}$
(2) $\alpha \neq ik$, $k \in \mathbf{Z}$, and under this condition the solution is unique
(3) The condition is $g_{-2} \propto \big(\exp(-2ix), g\big) = 0$. The general solution is

$$f(x) = \sum_{k \neq -2} \frac{g_k}{i(k+2)} \exp(ikx) + c \exp(-2ix), \qquad c = \text{const}$$

(4) $\|T^{-1}\| = \sup_k \frac{1}{|ik+\alpha|}$, then resp. $\|T^{-1}\| = 1,\ 1/2,\ 1,\ 2$

(1.92)
(1)

$$(g, Tf) = \int_0^\pi g^*(x) f''(x)\, dx = \int_0^\pi g''^*(x) f(x)\, dx + \big[g^*(\pi) f'(\pi) - g^*(0) f'(0)\big] - \ldots =$$

(Tg, f), thanks to the boundary conditions.
(2) With $f(x) = \sum f_n \sin nx,\ g = \sum g_n \sin nx$ one obtains $f_n = -g_n/n^2$ for $n = 1, 2, \ldots$
(3) The comparison with the next problem shows that, changing the domain, one obtains *different* Hermitian operators!

(1.93)
(2) One has $-4n^2 f_n = g_n$, which can be solved only if $g_0 = (e_0, g) = 0$ $\big($i.e. $g(x)$ must be zero mean-valued$\big)$; in this case a solution exists not unique (f_0 remains arbitrary, and any constant can be added)

(1.94)
(1) The solution is given by $f_n = g_n/(n^2+1),\ n = 1, 2, \ldots$ with usual notations.
(2) $\|S\| = \sup_n \frac{1}{n^2+1} = \frac{1}{2}$
(3) Yes, because S is a bounded operator
(4) $\|f - f_N\|^2 = \sum_{n>N} \dfrac{|g_n|^2}{(1+n^2)^2} \le \dfrac{1}{\big(1+(N+1)^2\big)^2}\|g - g_N\|^2$

(1.95)
(1) Eigenfunctions $\exp(inx),\ n \in \mathbf{Z}$; eigenvalues doubly degenerate if $n \neq 0, \ldots$
(2) Ker T is the 2-dimensional subspace generated by $\exp(\pm ix)$
(3) The solution exists if the Fourier coefficients $g_{\pm 1} \propto \big(\exp(\pm ix), g\big) = 0$. In this case, the equation admits ∞^2 solutions (the coefficients $f_{\pm 1}$ remain arbitrary)

$g(x) = \cos^4 x$ has no components $\propto \exp(\pm ix)$, then the equation admits solution, the contrary if $g(x) = \cos^3 x$
(4) T_α admits bounded inverse if $\alpha \neq n^2$
(5) $\|T_\alpha^{-1}\| = \sup_n \left|\frac{1}{-n^2+1+i\alpha}\right| = \frac{1}{|a|}$

(1.96)
(1) The same calculations as in Problem 1.92q.(1) show that T is Hermitian in the dense domain defined by the given boundary conditions. Its eigenfunctions $\cos(nx/2),\ n = 0, 1, 2, \ldots,$ are an orthogonal complete system in $L^2(0, 2\pi)$ exactly as $\cos nx$ are complete in $L^2(0, \pi)$. Ker T is the 1-dimensional subspace of the constants and Ran T its orthogonal complementary subspace
(2) $g(x)$ must be zero mean-valued; in this case, written $g = \sum_{n=1}^{\infty} g_n \cos(nx/2)$, the most general solution $f(x)$ is, where f_0 is an arbitrary constant,

$$f(x) = f_0 - 4\sum_{n=1}^{\infty} \frac{1}{n^2} g_n \cos\frac{nx}{2}$$

(3) $\left|\dfrac{df}{dx}\right| \le 2\sum_{n\neq 0} \dfrac{1}{n}|g_n|$; the r.h.s. can be viewed as a scalar product between two sequences $\in \ell^2$, and observing that $\sum |g_n|^2 = \frac{1}{\pi}\|g\|^2$, one obtains

$$\left|\frac{df}{dx}\right| \le K\|g\|_{L^2} \text{ with } K = \frac{2}{\sqrt{\pi}}\Big(\sum_{n\neq 0}\frac{1}{n^2}\Big)^{1/2} = \Big(\frac{2\pi}{3}\Big)^{1/2}$$

(1.97)
(1) $C = 1$
(2) $f_\alpha(x)$ converges to $h(x)$ in the norm L^2 and also uniformly. Notice that $f_\alpha(x)$ and $h(x)$ are continuous functions
(3) As in the previous problem, one has $|f(x)| \le \sum_{n\ge1} |\varphi_n|/n^2$, this series is now intended as the scalar product of the two functions: $G(x) = \frac{2}{\pi}\sum \sin nx/n^2$ and $\Phi(x)$

(1.98)
(1) $u_n(x) = \exp(cx/2)\sin nx$; $n = 1, 2, \dots$; $\lambda = n^2 + (c^2/4)$. The eigenfunctions are a complete set.
(2) $\|T^{-1}\| = 1/\big(1 + (c^2/4)\big)$

(1.99)
(1) $u_n(x) = \exp(in\pi x^2)$, $n \in \mathbf{Z}$
(2) $u_n(x)$ given in (1) are orthonormal with respect to the scalar product $(f, g)_\rho$; they are a complete set: impose the usual completeness condition $(u_m, h)_\rho = 0$, $\forall n$, then put $y = x^2 \dots$
(3) $c_n = (u_n, f)_\rho/(u_n, u_n)_\rho = (u_n, f)_\rho = \dots$
(4) $\sum_n |c_n|^2 = (f, f)_\rho = 2$
(5) No periodic. At the points $x = \pm 1/2,\ \pm\sqrt{9/4 + 2m}$, $m = 1, 2, \dots$

(1.100)
(1) *(a)* $T^+ = -T - I$
(2) *(b)* $c(\alpha) = \exp(\alpha/2)$
(3) One has $B = A$ and $\widetilde{B} = \widetilde{A}$; these operators are the infinitesimals Lie generators of the transformations defined in (2). According to Stone theorem, the generator of a unitary transformation is anti-Hermitian; see also Problem 4.18, q.(1) (a)

(1.101)
(1) Eigenvectors $\{e_{nm} = \sin(nx)\sin(my);\ n, m = 1, 2, \dots\}$, $\lambda = -(n^2 + m^2)$, degeneracy 1 or 2 (exceptionally 3: e.g., $\lambda = -50$ with eigenfunctions $\sin 5x \sin 5y$, $\sin 7x \sin y$, $\sin x \sin 7y$, see also Problem 4.7)
(2) $\|T^{-1}\| = 1/2$
(3) The (unique) solution is given by $f_{nm} = -g_{nm}/(n^2 + m^2)$ with clear notations
(4) The property $g(x, y) = g(y, x)$ is equivalent to $g_{nm} = g_{mn}$ then also $f_{nm} = f_{mn}$
(5) The common eigenfunctions are $\{e_{nm} \pm e_{mn}\}$, which provide an orthogonal complete system

(1.102)
(1) α, β purely imaginary
(2) $\lambda = \lambda_{k_1,k_2} = (ik_1 - \alpha)(ik_2 - \beta)$, the eigenvectors

$$u_{k_1,k_2} = \exp(ik_1x)\exp(ik_2y)\,, \qquad k_1, k_2 \in \mathbf{Z}$$

are an orthogonal complete system for $L^2(Q)$
(3) $\mathtt{Ker}\, T \neq \{0\}$ if $\alpha = ik_1$ or $\beta = ik_2$ (or both, clearly)
(4) The solution exists (not unique) if the Fourier coefficients of $g(x, y)$ with respect to the orthogonal complete system obtained in (2) satisfy $g_{1,-1} = g_{-1,1} = 0$. The proposed equation is solved by $f(x, y) = -(1/5)\sin 2(x + y) + c_1 \sin(x - y) + c_2 \cos(x - y)$, where c_1, c_2 are arbitrary constants

(1.103)
(1) $\lambda = \lambda_{k_1,k_2} = i(k_1 + \alpha k_2)$, the eigenvectors $u_{k_1,k_2} = \exp(ik_1x)\exp(ik_2y),\ k_1, k_2 \in \mathbf{Z}$, are an orthogonal complete system for $L^2(Q)$
(2) *(a)* In $\mathtt{Ker}\, T$ one has $k_1 = -k_2$; an orthogonal complete system for $\mathtt{Ker}\, T$ is $v_k = \exp\big(ik(x - y)\big),\ k \in \mathbf{Z}$; *(b)* If $\alpha = \sqrt{2}$ then $k_1 = k_2 = 0$, dim of $\mathtt{Ker}\, T = 1$
(3) $u(x, \pi) = 0$ if $0 < x < \pi$; and $u(x, \pi) = 1$ if $\pi < x < 2\pi$
(4) $u(x, y) = \varphi(x - y)$ where now φ is periodically prolonged with period 2π

(1.104)
(1)*(a)* The representative vector of the first functional is $(\underbrace{1, 1, \ldots, 1}_{N}, 0, 0, \ldots)$ and $\|\Phi\| = \sqrt{N}$; this confirms that the second functional is unbounded and not closed
(1) *(b)* $c_n \in \ell^2$
(3)*(a)* The representative vector of Φ is $\exp(|x|)$
(b) The functionals are unbounded and not closed; the same for the functional in (5)
(4) $\alpha > -1/2$; $\beta < -1/2$; no γ
(5) E.g., $g_n(x) = \begin{cases} 1 & \text{for } |x - x_0| > 1/n \\ n^{1/3} & \text{for } |x - x_0| < 1/n \end{cases}$, then $\|g_n - f\| = 2n^{-1/3} \to 0$

(1.105)
(1) Strong convergence to 0, indeed, $|\Phi_n(v)| = |(e_n, v)| \to 0$; whereas $\|\Phi_n\| = \|e_n\| = 1$ (this is a restatement of a well-known property of orthonormal sets $\{e_n\}$)
(2) $S_v = (w, v)$ (this is a restatement of the Parseval identity)
(3) $S_g = \frac{\pi}{4}(f_1, g) = \frac{\pi}{4}\int_0^{\pi}\big(g(x) - g(-x)\big)\,dx$

(1.106)
(1) $\|\Phi\| = 1$
(2) Let $f(x) \in L^2(0, 1)$ and $\widetilde{f}(x) \in L^2(0, 1)$ be a continuous function "approximating" $f(x)$, i.e., $\|f - \widetilde{f}\| < \varepsilon$. Then $|\Phi(f)| \le |\Phi(f - \widetilde{f})| + |\Phi(\widetilde{f})|$; apply now the mean-value theorem: $\Phi(\widetilde{f}) = \frac{1}{\sqrt{a}}\int_0^a \widetilde{f}(x)\,dx = \sqrt{a}\,f(\xi)$ with $0 < \xi < 1$

(1.107)
As in the problem above, given $f(x)$, let $\widetilde{f}(x)$ be a continuous function with compact support K "approximating" $f(x)$, then $\Phi(\widetilde{f}) = \frac{1}{\sqrt{a}}\int_0^a \widetilde{f}(x)\,dx = \frac{1}{\sqrt{a}}\int_{x \in K}\widetilde{f}(x)\,dx$, etc.

(1.108)
(1) $u(x,t) = \sum_{n=1}^{+\infty} a_n \exp(-n^2 t) \sin nx$, where $a_n = \frac{2}{\pi}(\sin nx, f)$
(2)

$$\|u(x,t)\|^2 = (\pi/2) \sum_{n=1}^{+\infty} |f_n|^2 \exp(-2n^2 t) \leq \exp(-2n_1^2 t) \|f(x)\|^2$$

where f_n are the coefficients of the Fourier expansion of $f(x)$ and n_1 the first nonzero coefficient of this expansion
(3) The coefficients $\exp(-n^2 t)$ "compensate" the terms n^k coming from differentiation, and ensure convergence of the series for any $k > 0$.

(1.109)
(1) The eigenvectors are $\sin nx$
(2) $\|E_t\|^2 = \sup_n \frac{\sum_n |a_n|^2 \exp(-2n^2 t)}{\sum_n |a_n|^2} = \exp(-2t)$
(3)

$$\|E_t - I\|^2 = \sup_n \frac{\sum_n |f_n|^2 \big(\exp(-n^2 t) - 1\big)^2}{\sum |f_n|^2} = 1$$

In strong sense:

$$\sum_{n=1}^{\infty} |f_n|^2 \big(\exp(-n^2 t) - 1\big)^2 \leq \sum_{n=1}^{N} |f_n|^2 + \sum_{n=N+1}^{\infty} |f_n|^2 \to 0$$

Indeed, given $\varepsilon > 0$, there is N such that

$$\sum_{n>N} |f_n|^2 \big(\exp(-n^2 t) - 1\big)^2 \leq \sum_{n>N} |f_n|^2 < \varepsilon$$

on the other hand, being $\exp(-n^2 t) \to 1$ as $t \to 0^+$, there is t such that

$$\sum_{n=1}^{N} |f_n|^2 \big(\exp(-n^2 t) - 1\big)^2 < \varepsilon$$

then ...

(1.110)
(2) $u(x,t)$ tends, in the $L^2(-\pi,\pi)$ norm, to the mean value a_0 of $f(x)$
(3) $\int_{-\pi}^{+\pi} f(x)\,dx = 0$ means $a_0 = 0$ and this property is preserved for all t
(4) $E_t \to P_0$ in norm as $t \to +\infty$, where P_0 is the projection on the constants; $E_t \to I$ strongly as $t \to 0^+$, see previous problem

(1.111)
The equation for a_n is $\dot a_n = -n^2 a_n + F_n$, the solution is

$$u(x,t) = \sum_{n\in\mathbf{Z}} \left(\left(f_n - \frac{F_n}{n^2} \right) \exp(-n^2 t) + \frac{F_n}{n^2} \right) \exp(inx)$$

(1.112)
(1) $\displaystyle\frac{d}{dt}\sum_{n=1}^{+\infty} a_n(t)\, e_n = \sum_{n=1}^{+\infty} -n^2 a_n(t)\, e_n$ then $\dot a_n = -n^2 a_n$, etc., exactly as in Problems 1.108-1.110

(1.113)
In general, writing $v(t) = \sum_{n\in\mathbf{Z}} a_n(t)\, e_n$, one obtains $\dot a_0 = a_0$; $\dot a_n = a_{-n}$, then

$$v(t) = c_0 e_0 + \sum_{n\ge 1} \big(b_n \exp(t) + c_n \exp(-t)\big) e_n + \big(b_n \exp(t) - c_n \exp(-t)\big) e_{-n}$$

If $v(0) = e_0$ then $v(t) = e_0 \exp(t)$. If $v(0) = e_n$ then $v(t) = e_1 \cosh t + e_{-1} \sinh t$

(1.114)
In general, writing $v(t) = \sum_{n\in\mathbf{Z}} a_n(t)\, e_n$, one obtains $\dot a_0 = 0$; $\dot a_n = -n\, a_{-n}$, then

$$v(t) = c_0 e_0 + \sum_{n\ge 1} \big(b_n \exp(int) + c_n \exp(-int)\big) e_n + i\big(- b_n \exp(int) + c_n \exp(-int)\big) e_{-n}$$

If $v(0) = e_0$ then $v(t) = e_0$. If $v(0) = e_n$ then $v(t) = e_n \cos nt + e_{-n} \sin nt$

(1.115)
(1) $u_n(x) = \exp\big(ix(n + \frac{1}{2})\big)$, $n \in \mathbf{Z}$; the eigenvectors are an orthogonal and complete set (easy to check); the eigenvalues are $\lambda_n = i(n + \frac{1}{2})$
(2) $a_n(t) = a_n \exp\big(it(n + \frac{1}{2})\big)$, where $f_0(x) = \sum_{n\in\mathbf{Z}} a_n \exp\big(ix(n + \frac{1}{2})\big)/\sqrt{2\pi}$
(3) Period $T = 4\pi$; notice that the solution is simply $f_0(x + t)$
(4) E_t is unitary: this follows from (1) or from (3)
(5) No convergence in norm:
$\|E_t - I\| = \sup_n \big|\exp\big(it(n + \frac{1}{2})\big) - 1\big| = 2$; this $\sup_n$ turns out to be independent of t.
Strong convergence: $\|(E_t - I) f(x,0)\|^2 = \sum_{n\in\mathbf{Z}} |a_n|^2 |\exp(i(n + \frac{1}{2})t) - 1|^2 \to 0$, using, e.g., the same argument as in Problem 1.109, q.(3)

(1.116)
(1) and (2)(a) Notice that $\widetilde{A}^2 = \widetilde{B}^2 = 0$, then putting $A = I + \widetilde{A}$:
$$\exp(At) = \exp(t)\exp(\widetilde{A}t) = \exp(t)(I + \widetilde{A}t) = \begin{pmatrix} \exp(t) & 0 \\ t\exp(t) & \exp(t) \end{pmatrix}$$

(2) *(b)* $u(t) = \big(\exp(t)\big)(e_1 + t\,e_2)$ and $u(t) = \exp(t)(e_2 + e_3 + te_4)$
(c) Writing $u_0 \equiv (a_1, a_2, a_3, \ldots)$, one has $u(t) \equiv (a_1,\ ta_1 + a_2,\ a_3,\ ta_3 + a_4, \ldots)$

(1.117)
(1) Only if $f(x) = 0$
(2) If, e.g.., $f(x) = \sin nx$, and then for any finite linear combination thereof
(3) $f(x) = \sum_{n\neq 0}(1/n^2)\sin nx$
(4) No: e.g.., $f_N(x) = (1/N)\sin Nx$

(1.118)

$$(2)\ u_1(x,0) = (1/2)|\sin 2x|\,; \quad u(x,\pi) = \begin{cases} 0 & \text{for } 0 \le \pi/2 \le \pi \\ -\sin 2x & \text{for } \pi/2 \le x \le \pi \end{cases}$$

(1.119)
(2) Strong convergence to $-I$: $\|(I + T_a)f\|^2 = \int_{x>a}|f(x)|^2\,dx \to 0$.
Instead: $\|I + T_a\| = 1$ $\big($choose a $f(x)$ with support in $x > a\big)$
(3) Let $a > b$, then $T_aT_b = P_b - P_{ba} + P_a$, where P_b projects on $H_b = L^2(-\infty, b)$; P_{ba} projects on $H_{ba} = L^2(b,a)$; P_a projects on $H_a = L^2(a, +\infty)$
(4) Let $\{e_n\}_b$, $\{e_m\}_{ba}$, $\{e_p\}_a$ be any orthogonal complete sets in H_b, H_{ba}, H_a resp., they are simultaneous eigenvectors of T_a and T_b, and their union is an orthogonal complete set in $L^2(\mathbf{R})$

(1.120)
(1) $T(v_n) = 0$ for even n; $T(v_0) = 1$; $T(v_n) = \frac{2i}{\pi n}\exp(inx)$ for odd n.
(2) See Problem 1.25
(3) Yes, thanks to the properties of its eigenvalues
(4) $\|T^N\| = \sup_{n=1,3,5,\ldots}\{1, \big(2/n\pi\big)^N\} = 1$
(5) T^N is norm-convergent to P_0, i.e., the projection on the 1-dimensional subspace of constants: $\|T^N - P_0\| = \sup_{n=1,3,5,\ldots}\big(2/n\pi\big)^N = (2/\pi)^N \to 0$

(1.121)
(1) $\|T\| = \max\{|\alpha + \beta|, |\alpha - \beta|\}$
(2) The even functions $f^+(x)$ $\big($resp., the odd functions $f^-(x)\big)$ are eigenfunctions with eigenvalue $\alpha + \beta$ (resp. $\alpha - \beta$).
The operator can then be written as $Tf = (\alpha + \beta)f^+ + (\alpha - \beta)f^-$
(3) $|\alpha + \beta| = |\alpha - \beta| = 1\,;\ \beta = \pm i/\sqrt{2}$
(4) $Tf = f^+(x) + (1/3)f^-(x)$, then T^n is norm convergent to the projection on the even functions in $L^2(\mathbf{R})$

(1.122)
(1) T_n and S_n are Hermitian, indeed, e.g.., $T_n^+ = P_n\,\frac{x}{1+x^2} = \frac{x}{1+x^2}P_n = T_n$.
$\|T_n\| = 1/2,\ \|S_n\| = 1$
(2) $g(x)$ must have support in $[0, n]$ and have some "good" behavior at $x = 0$ $\big($cf. Problems 1.6, q.(2), and 1.82, q.(3)$\big)$. E.g., $g(x) = 1$ for $0 \le x \le n$ and $= 0$ for $x > n$

does not belong to $\mathtt{Ran}\, T_n$, which however is dense in the subspace $L^2(0, n)$. With the above assumptions on $g(x)$, the solution is unique.
(b) $\mathtt{Ran}\, T_n$ is not a Hilbert subspace; one has $H = \mathtt{Ker}\, T_n \oplus \overline{\mathtt{Ran}\, T_n}$, where $\overline{\mathtt{Ran}\, T_n}$ is the closure of $\mathtt{Ran}\, T_n$
(3) $P_n f(x) = \frac{\sin \pi x}{x}(1 + x^2)P_n g(x)$: there is no singularity at $x = 0$. The solution $f(x)$ exists $\big($not unique, due to the arbitrariness of $f(x)$ for $x > n\big)$ for any $g \in H$
(4) $\|(I - P_n)\frac{x}{1+x^2}\| = \sup_{x>n} \frac{x}{1+x^2} \to 0$, then T_n is norm-convergent to the multiplication operator $(x/1 + x^2)\, I$.

$\|(I - P_n) \sin \pi x\, f(x)\|^2 \le \int_{x>n} |f(x)|^2\, dx \to 0$, but $\|(I - P_n) \sin \pi x\| = 1$ and S_n converges strongly, not in norm, to the multiplication operator $\sin \pi x\, I$

(1.123)
(1) $T^4 = I$
(2) The eigenvalues are:
$\lambda = 1$ with eigenspace H_1 generated by e_n where $n = 4m,\ m \in \mathbf{Z}$;
$\lambda = i$ with eigenspace H_i generated by e_n where $n = 4m + 1$;
$\lambda = -1$ with eigenspace H_{-1} generated by e_n where $n = 4m + 2$;
$\lambda = -i$ with eigenspace H_{-i} generated by e_n where $n = 4m + 3$.
$|\sin 2x| \in H_1$ has period $\pi/2$ and is eigenvector of T
(3) $\mathtt{Ker}\, S = H_{-1}$, $\mathtt{Ran}\, S$ is the complementary subspace to $\mathtt{Ker}\, S$

(1.124)
(1) Indeed, T is a combination of two translations
(2) The eigenvalues are $1,\ 0,\ -1$
(3) $T = P_+ - P_-$ where P_+ projects on the subspace generated by e_k with $k = 4m + 1, m \in \mathbf{Z}$, etc.
(4) $\|T\| = \max\{1,\ 0,\ -1\} = 1$; $\|T + (1 + 2i)I\| = \max\{|2 + 2i|,\ |1 + 2i|,\ |2i|\} = 2\sqrt{2}$; $\|\big(T + (1 + 2i)I\big)^{-1}\| = 1/\min\{\ldots\} = 1/2$

(1.125)
(1) An orthonormal complete system for $\mathtt{Ran}\, T$ is $e_0 = 1/\sqrt{2},\ e_1 = \sqrt{3/2}\, x$
(2) With $A = 1/2$, $B = 3/2$, then $T_{AB} = P_0 + P_1$, with clear notations, is a projection, see (1), cf. Problem 1.45, q.(1)
(3) Apart from the eigenvalue $\lambda = 0$, T admits the eigenvalues 2 and 2/3 with eigenvectors resp. e_0 and e_1
(4) $Tf = 2e_0(e_0, f) + (2/3)e_1(e_1, f)$ then

$$\|T\|^2 = 4|f_0|^2 + (4/9)|f_1|^2 \le 4(|f_0|^2 + |f_1|^2) \le 4\|f\|^2$$

and $\|T\| = 2$. Notice that in $Tf = \int_{-1}^{1} f\, dx + x(x, f)$ the term $\int_{-1}^{1} f\, dx = (1, f) = \sqrt{2}(e_0, f)$ is to be *consistently* intended as a *constant function* (not as a functional), then e.g., $\|1\| = \sqrt{2}$ and

$$\|\int_{-1}^{1} f\, dx\| = \|1(1, f)\| = \sqrt{2}|(1, f)| \le 2\|f\|$$

The result $\|T\| = 2$ is confirmed by $\|T\| = \max|eigenvalues| = 2$, thanks to the fact that the eigenvectors of T provide an orthogonal complete system
(5) An orthonormal complete set for $\mathrm{Ran}\, T$ is $e_0 = 1/\sqrt{2}$ and $e' = \sqrt{5/8}(3x^2 - 1)$. Evaluate $T_{AB}(1)$ .... For no $A,\ B$ both nonzero, T_{AB} is a projection

(1.126)
(1)(b) The eigenvalues are: α, with eigenvectors the zero mean-valued functions, and $\alpha + 2\beta\pi$ with eigenvector the 1-dimensional subspace of constant functions. The eigenvectors provide an orthogonal complete system for $L^2(-\pi, \pi)$. Then $\|T\| = \max\{|\alpha|,\ |\alpha + 2\pi\beta|\}$. See previous problem for what concerns the term $\int_{-\pi}^{\pi} f(y)\, dy$
(2) $\beta = -1/(2\pi)$, see (1)
(3) T_n converges weakly to the operator αI: $\big(g, (T_n - \alpha I)f\big) = 2\pi\beta(g, e_n)(e_0, f)$ $\to\ 0$, where $e_n = \exp(inx)/\sqrt{2\pi}$; no strong convergence:

$$\|T_n f - \alpha f\| = 2\pi|\beta|\,\|e_n\|\,|(e_0, f)| = 2\pi|\beta|\,|f_0|$$

(1.127)
(1)(a) The eigenfunctions of D are $\exp(ikx),\ k \in \mathbf{Z}$, with non-degenerate eigenvalues
(b) $T_a D = DT_a = (1/2a)\big(f(x+a) - f(x-a)\big)$
(2) Thanks to (1), the eigenfunctions of D are eigenfunctions of T_a: $D(T_a v) = T_a D v = \lambda(T_a v)$, then $T_a v = \mu v$. The eigenvalues μ of T_a are $\lambda_k = \sin ka/(ka),\ k \in \mathbf{Z}$, with eigenfunctions $\exp(ikx)$, doubly degenerate (apart from the case $k = 0$) for "generic" a; if, e.g.., $a = \pi/2$ then ...
(3) T_a is Hermitian, because its eigenvalues are real and its eigenfunctions are an orthogonal complete system; T_a is compact, thanks to the properties of its eigenvalues
(4) $T_a f(x) = \sum_{k \in \mathbf{Z}} c_k \frac{\sin ka}{ka} \exp(ikx)$
(a) $T_\pi f(x) = c_0$, the mean value of $f(x)$, and $\mathrm{Ran}\, T_\pi$ is 1-dimensional
(b) m must be odd
(c) $\mathrm{Ran}\, T_a$ contains the complete set $\{\exp(ikx)\}$, therefore is dense in H, but, e.g., $g(x) = \sum_{k \neq 0}(1/k)\exp(ikx) \notin \mathrm{Ran}\, T_a$. Then, T_a^{-1} exists unbounded
(5) $\|T_a\| = \sup_k |\lambda_k| = 1$. $T_a \to I$ strongly, it is possible to use an argument similar to that seen in Problem 1.109; no norm convergence: $\|T_a - I\| = \sup_k \left|\dfrac{\sin ka}{ka} - 1\right| = 1$

(1.128)
(2) The eigenvalues are $\lambda_n = \exp(ina) - 1$, infinitely degenerate if $a = \pi/2$, not degenerate if $a = 1$
(3) $TT^+ f(x) = T^+ T f(x) = 2f(x) - f(x-a) - f(x+a)$.
$\|T_a\| = \sup|\text{eigenvalues}| = 2$
(4) For any a; indeed $\exp(ina) - 1 \neq i,\ \forall n$.
$\|(T_a - iI)^{-1}\| = 1$ if $a = \pi/2$. If $a = 1$: $\|(T_a - iI)^{-1}\| = \sup_n \dfrac{1}{|\exp(ian) - 1 - i|} =$
$= \big(\text{distance of}\quad 1+i\quad \text{from the circumference of radius 1 centered in zero}\big)^{-1} = \frac{1}{\sqrt{2}-1}$

(1.129)
(1) $\|T\| = 1,\ T = T^{-1}$.
(2) T is unitary: it is invertible and $\|Tf\| = \|f\|$
(3) For any $f(x)$ one has that $f(x) \pm Tf(x)$ are eigenfunctions

(1.130)
(1) T is unbounded
(2) Only $f_2(x)$ and $f_4(x)$ belong to the domain of T
(3) E.g., $f_2(x) \pm Tf_2(x)$ are eigenfunctions of T

(1.131)
(1) *(a)* With $n, m \neq 0$, one has $(e_n, T\, e_m) = \delta_{nm}/im$; and $(e_0, T\, e_m) = (i/m)\delta_{m0}$, etc.
(b) T is bounded
(c) Differentiating $Tf = \lambda f$ one sees that any "possible" eigenfunction must have the form $f(x) = c\ \exp(x/\lambda)$, but inserting into $Tf = \lambda f$ one concludes that there are no eigenfunctions.
(2) $f(x) = c\ \exp(x/\lambda)$ is eigenfunction for any λ with $\mathrm{Re}\,\lambda > 0$
(3) No eigenfunctions

(1.132)
(1) The 4 subspaces are mutually orthogonal. The integrals give resp. $f^{(1)}(x)$, 0, 0
(2) $T^2 = T$. $\mathrm{Ran}\,T = H^{(0)} \oplus H^{(1)}$, $\mathrm{Ker}\,T = H^{(2)} \oplus \widetilde{H}$.
$\|T\,f\|^2 = |a_0|^2 + \|f^{(1)}\|^2 \le \|f\|^2$ then $\|T\| \le 1$, but choosing, e.g., $f = 1$, one has $\|T\| = 1$
(3) $T_k^2 = 0$. $\mathrm{Ran}\,T_k = \mathrm{Ker}\,T_k = H^{(0)} \oplus H^{(1)}$
(4) $\|T_k\| = 1$; strong convergence to 0: $\|T_k f(x, y)\|^2 = |a_{0k}|^2 + \sum_{n\neq 0} |a_{nk}|^2 \to 0$

Problems of Chap. 2

(2.1)
(1)

$$(1+i)^{100} = \left(\sqrt{2}\exp(i\pi/4)\right)^{100} = 2^{50}\exp(25\pi i) = -2^{50}$$

(2) All these equations admit infinite solutions; e.g..: $\cos z = 3$, with $t = \exp(iz)$, becomes $t + t^{-1} - 6 = 0$, then $\exp(iz) = 3 \pm 2\sqrt{2}$ and two families of solutions are found:

$$z_n^+ = -i\log(3+2\sqrt{2}) + 2n\pi, \text{ and } z_n^- = -i\log(2\sqrt{2}-3) + \pi + 2n\pi;\ n \in \mathbf{Z}$$

(2.2)
$z^2\sin(1/z)$ tends to zero if z moves along the real axis $z = x$, but along the imaginary axis $z = iy$ .... Notice that the function is zero at $z = 1/n\pi,\ n = \pm 1, \pm 2, \ldots$, then

$z = 0$ is an accumulation point of zeroes, and this is enough to conclude that it is an essential singularity: indeed, it cannot be a point of analyticity (or a removable singularity), because this would imply that it is a non-isolated zero (which in turn would imply $f(z) \equiv 0$!); on the other hand, it cannot be a pole, because this would imply $|f(z)| \to +\infty$ for $z \to 0$.

For the second function: $\exp(-1/z^4)$ tends to zero if z tends to zero along the real and imaginary axes, but along the line $z = |z| \exp(i\pi/4)$ ..., and has no limit choosing $z = |z| \exp(i\pi/8)$. Similar arguments for the other functions

(2.3)

(1) $f(z) = c_0 + \dfrac{-3i}{z-i} + \dfrac{c_{-2}}{(z-i)^2}$ plus $\sum_{n=1}^{\infty} c_n z^n$ if $f(z)$ is not analytic at $z = \infty$

(2) $f(z) = c_0 + \dfrac{1}{z-1} + \dfrac{c_{-2}}{(z-1)^2} + \dfrac{2}{(z-1)^3} + \sum\limits_{n=1}^{\infty} c_n z^n$

(i) $f''(z)$ has a pole of order 5 at $z = 1$ and zero residue. It can be analytic at $z = \infty$ if $f(z)$ has (at most) a double pole at ∞, i.e., if $c_3 = c_4 = \ldots = 0$

(ii) The primitive has a logarithmic singularity at $z = 1$

(2.4)

(1) From the property $u_{xx} + u_{yy} = 0$, one has $a_{xx} + b_{yy} = 0$ which implies $a_{xx} = -b_{yy} =$ const, then $f(z) = c_0 + c_1 z + c_2 z^2$, $c_i =$const $\in \mathbf{C}$.

(2) Using the same procedure, $f(z) = c_0 \exp(c_1 z)$

(2.5)

(a) The Taylor series converges in $|z| < 1$; *(b)* For the evaluation of the expansion, it can be useful to write $f(z) = \dfrac{1}{(1-2i) - (z-2i)} = \dfrac{1}{(1-2i)} \dfrac{1}{1 - \frac{z-2i}{1-2i}} = \ldots;$ the series converges in $|z - 2i| < |1 - 2i| = \sqrt{5}$;

(c) $f(z) = \dfrac{1}{z(\frac{1}{z} - 1)} = -\dfrac{1}{z} \sum\limits_{n=0}^{\infty} \dfrac{1}{z^n}$ this series converges in $|z| > 1$;

(d) no need of expansion in powers of $(z-1)^n$: $f(z)$ is itself a Taylor-Laurent series containing the only term $-1/(z-1)$

(2.6)

(1) $\sum\limits_{n=1}^{\infty} n z^n = z \sum\limits_{n=1}^{\infty} n z^{n-1} = z \dfrac{d}{dz}\Big(\sum\limits_{n=0}^{\infty} z^n\Big) = z \dfrac{d}{dz}\Big(\dfrac{1}{1-z}\Big) = \dfrac{z}{(1-z)^2}$

(2) $R = 1$

(3) There is convergence in the annulus $2 < |z| < 3$

(2.7)

The series converges if $|\exp(-z)| = \exp(-x) < 1$, or $x > 0$. $S(z) = \frac{1}{1-\exp(-z)}$.
The point $z = \infty$ is an accumulation point of simple poles at $z = 2n\pi i$, $n \in \mathbf{Z}$

(2.8)
Essential singularity at $z = \infty$; $\mathrm{R}(\infty) = 0$

(2.9)
Expand $f(z) = f(z_0) + (z - z_0)f'(z_0) + \ldots = (z - z_0)f'(z_0) + \ldots$, the same for $g(z)$

(2.10)
$z = \pm i\pi$ are removable singularities; $z = \pm 3i\pi,\ \pm 5i\pi, \ldots$ are simple poles; $R = 3\pi$

(2.11)
$f_1(z)$ has simple poles at $z = n\pi,\ n \in \mathbf{Z}$; $z = \infty$ is an accumulation point of simple poles. $z = 0$ is an accumulation point of simple poles (at $z = 1/n\pi$) for $f_2(z)$

(2.12)
(1) $a_{-1} = 1;\ a_0 = a_2 = 0;\ a_1 = 1/6$
(2) $f_2(z) = \frac{1}{6} + \frac{z^2}{120} + \ldots$

(2.13)
$f_1(z)$ has a double pole at $z = 0$ and an essential singularity at $z = \infty$
$f_2(z) = \frac{-z^3}{30} + \ldots$ has only an essential singularity at $z = \infty$
(recall the Taylor expansions of $\sin z,\ \cos z$)

(2.14)
Pole at $z = 0$ of order $n - 8$ if $n > 8$, etc. (use the Taylor expansion $\exp(\pm z^2) = 1 \pm z^2 + z^4/2! + \ldots$); essential singularity at $z = \infty$

(2.15)
In the cases where branch points are present, the cut line will be put along the positive real axis, and $f_\pm(x)$ will denote the limit of the function $f(z)$ when z approaches the "upper" and resp. the "lower" margin of the cut.

$f_1(z)$: branch points at $z = 0$ and $z = \infty$; the discontinuity $f_+ - f_-$ is $2 \sin\sqrt{x}$;

$f_2(z)$: no branch points, indeed $f_+ = f_- = \sin^2(\sqrt{x})$, essential singularity at $z = \infty$; the same for $f_3(z)$ and $f_5(z)$

$f_4(z)$ and $f_6(z)$: branch points at $z = 0$ and $z = \infty$;

$f_7(z)$ and $f_8(z)$: branch points at $z = 0$ and $z = \infty$;

$f_9(z)$: essential singularity at $z = 0$ analytic for all $z \neq 0$

(2.16)

$f_1(z)$: branch points at $z = 0$ and $z = 1$; putting the cut along the positive real axis from $x = 0$ and $x = 1$, the discontinuity is (with the notations as in the previous problem) $f_+ - f_- = -2\pi i$;

$f_2(z)$: branch points at $z = 0,\ 1,\ \infty$; with the cut along the positive real axis, the discontinuity is $-2\pi i$ in $0 < x < 1$ and $-4\pi i$ in $x > 1$;

$f_3(z)$: branch points at $z = 0$ and $z = 1$;
$f_4(z)$: branch points at $z = 0,\ 1,\ \infty$

Consider the function $f_5(z)$: $z = 0$ is a branch point, indeed, starting, e.g., at some real point x with $0 < x < 1$, and moving continuously the point counterclockwise along a closed path encircling $z = 0$, when the point reaches its initial position, the function $f_5(z)$ changes its value by a factor $\exp(4\pi i/3)$. A similar procedure around $z = 1$ shows that $f_5(z)$ changes by a factor $\exp(2\pi i/3)$. Taking finally a closed line around $z = \infty$, which is, in this case, a line encircling both $z = 0$ and $z = 1$, one sees that $f_5(z)$ would change by a factor $\exp(4\pi i/3) \cdot \exp(2\pi i/3) = 1$, then $z = \infty$ is not a branch point.

(2.17)
(1) $n \leq 5$; essential singularity at $z = \infty$
(2) Nothing changes: the branch points at $z = \pm 1$ do not prevent the expansion near $z = 0$, it is enough to place the cut in the real axis from $-\infty$ to 0 and from 1 to $+\infty$

(2.18)
(1) $\alpha = n\pi i\ ,\ n \in \mathbf{Z}$; essential singularity at $z = \infty$ for any α
(2) The points $z = \pm 1$ are branch points for all α, and $z = \infty$ an essential singularity

(2.19)
(1) Not true: choose, e.g., $z = -i$
(2) With the cut fixed in $x \geq 0$, one gets $\sqrt{-1} = i$. Then $f_1(z)$ has a simple pole at $z = -1$ with residue $\lim_{z \to -1} \dfrac{z+1}{\sqrt{z} - i} = 2i$; instead, $f_2(z)$ has no other singularity, in particular $f_2(-1) = 1/2i$; there is a simple pole at $z = -1$ in the second Riemann sheet of the function $\sqrt{z}$. Similarly, $\log(-1) = \pi i$. Then, $f_3(z)$ has a simple pole at $z = -1$ with residue $2\pi i$; finally, $f_4(z)$ has no other singularity, in particular $f_4(-1) = 1$; there is a simple pole at $z = -1$ in another Riemann sheet of the function $\log z$.

(2.20)
it is enough to place the cut in the real axis from $-\infty$ to -1 and from 1 to $+\infty$:
$f_1(z) = 1 \pm \frac{z^2}{2} - \frac{z^4}{8} + \ldots$; $\quad f_2(z) = 2z + \frac{2z^3}{3} + \ldots$; $\quad f_3(z) = \pm z^2 - \frac{z^4}{2} + \ldots$

(2.21)
(1–2) $\mathrm{R}(\infty) = -b_1$, where b_1 is the coefficient of $1/z$ in the Taylor-Laurent expansion of $f(z)$ "in the neighborhood of ∞"; one has $b_1 = 0$ in the present cases
(3) $\mathrm{R}(\infty) = 0$ ensures that the sum of the residues in the upper half plane is opposite to those in the lower half plane, the conclusion is reached considering the directions of the integration in the two cases

(2.22)
The first integral = $2\pi i\,\mathrm{R}(i) = \pi/2$; the second = $-2\pi i\,\mathrm{R}(-i) = i\pi/18$

(2.23)

The first integral $= \dfrac{2\pi i}{4!}\dfrac{d^4}{dz^4}(\sin z)\Big|_{z=1} = \dfrac{\pi i}{12}\sin(1)$;
second integral $= 2\pi i\big(\mathrm{R}(0)+\mathrm{R}(\pi)\big) = 2\pi i(1+\pi)$;
third integral $= 2\pi i\big(\mathrm{R}(\pi i)+\mathrm{R}(-\pi i)\big) = -8\pi i$

(2.24)

$\oint = 2\pi i\,\mathrm{R}(1/2) = \pi i/9$; $\oint = 2\pi i$

(2.25)

$$\int_0^{2\pi}\dots\,d\theta = \oint_{|z|=1}\frac{z^2+1}{z^2+4z+1}\frac{dz}{iz} = 2\pi i\big(\mathrm{R}(0)+\mathrm{R}(-2+\sqrt{3})\big) = \pi\big(2-4/\sqrt{3}\big)$$

The second and the third integrals are equal to, resp., $\pi\sqrt{2}$ and $-2\pi/(3\sqrt{3})$

(2.26)

$-2\pi a\exp(-a)$ with the sign $+$ in the first integrand, and 0 with the sign $-$
The second integral $= 2\pi i\,\mathrm{R}(i) = \pi/e$

(2.27)

The first integral $= \mathrm{Re}\int_{-\infty}^{+\infty}\frac{\exp(ix)}{1+x^2}\,dx = \mathrm{Re}\big(2\pi i\,\mathrm{R}(i)\big) = \pi/e$
The second integral $= -(2\pi/\sqrt{3})\exp(-\sqrt{3}/2)\sin(1/2)$

(2.28)

Using notations of Fig. 2.1, which refers to the first one of the integrals, one has

$$\int_{-R}^{-r} + \int_{-\gamma} + \int_r^R + \int_\Gamma = 2\pi i\,\mathrm{R}(i)$$

then

$$\lim_{r\to 0}\int_{-\infty}^{+\infty} = \mathrm{P}_0\int_{-\infty}^{+\infty} = 2\pi i\,\mathrm{R}(i) + \pi i\,\mathrm{R}(0) = \text{etc.}$$

where P_0 is the Cauchy principal part of the integral (with respect to the singular point at $x=0$). Taking the imaginary part of the result, the symbol P_0 is unnecessary. The result is $\pi(e-1)/e$.

The second and the third integrals are equal to π and $(\pi/2)\big(1+\exp(-\pi/2)\big)$.

For the last integral:

$\int\dots = \mathrm{Im}\Big[\pi i\big(\mathrm{R}_f(0)+\mathrm{R}_f(1)+\mathrm{R}_f(-1)\big)\Big]$, where $f = f(z) = \frac{\exp(i\pi z)}{z(1-z^2)}$. The result is 2π

(2.29)

$$\text{The first integral} = \mathrm{P}_0\int_{-\infty}^{+\infty}\frac{\exp(ix)}{2ix(x-i)}dx - \mathrm{P}_0\int_{-\infty}^{+\infty}\frac{\exp(-ix)}{2ix(x-i)}dx \equiv$$

$$\mathrm{Po}\int_{-\infty}^{+\infty} f^{(+)}dx - \mathrm{Po}\int_{-\infty}^{+\infty} f^{(-)}dx = 2\pi i\,\mathrm{R}_{f^{(+)}}(i) + \pi i\,\mathrm{R}_{f^{(+)}}(0) - \pi i\,\mathrm{R}_{f^{(-)}}(0) = \frac{\pi i(e-1)}{e}$$

(with clear notations). The second integral is equal to $2\pi(1-e)/e$

(2.30)
$\int = \pi i\,\mathrm{R}(0) = \pi$

(2.31)
The integrals are respectively equal to $\pi i\,\mathrm{R}(-1) = \pi(-1+i)/2$;
$-\pi i\,\mathrm{R}(0) - 2\pi i\,\mathrm{R}(-2i) = (\pi/2) - (\pi/9)e^{-2}$; $\pi i\,\mathrm{R}(0) + 2\pi i\,\mathrm{R}(i) = \pi i(4e^{-1} - 1)$

(2.32)
$\oint f(z)\,dz = 2\int_{-\infty}^{+\infty} f(x)\,dx = 2\pi i\,\mathrm{R}_f(\pi i/2)$; there is a simple pole in $z = \pi i/2$.
The integrals are resp. equal to $\pi^3/4$ and $\dfrac{\pi}{\cosh(a\pi/2)}$

(2.33)
$0 = \oint = \int_0^{+\infty} \ldots\, dx + \int_{+\infty}^{0} d\rho \exp(i\pi/4)\exp(i(\rho\exp(i\pi/4))^2)$ then
$\int_0^{+\infty} \exp(ix^2)dx = \frac{1+i}{\sqrt{2}}\int_0^{+\infty} d\rho\exp(-\rho^2) = \frac{1+i}{\sqrt{2}}\frac{\sqrt{\pi}}{2}$, etc.

(2.34)
(1) *(a)* $\mathrm{R}(-i) = \exp(3i\pi/4) = (-1+i)/\sqrt{2}$; *(b)* $\mathrm{R}(-i) = 3i\pi/2$; *(c)* $\mathrm{R}(-1) = i$;
(d) $\mathrm{R}(-i) = 4i/3$; *(e)* $\mathrm{R}(-i) = -i\sinh\pi$; *(f)* $\mathrm{R}(-i) = -\exp(-\pi)$
(2) *(a)* $\mathrm{R}(i) = \log|i-1| + i\,3\pi/4 + \log|i+1| + i\,\pi/4 = \log 2 + i\pi$; *(b)* $\mathrm{R}(i) = i\sqrt{2}$; *(c)* $\mathrm{R}(-1) = -i/2\sqrt{2}$; $\mathrm{R}(i) = (i/4)\sqrt[4]{2}\exp(3i\pi/8)$;
$\mathrm{R}(-i) = \sqrt{|-i-1|}/(4i)\exp(5i\pi/8) = -(i/4)\sqrt[4]{2}\exp(5i\pi/8)$
(3) *(a)* The cut lines of $\sqrt{z+1}$ and $\sqrt{z-1}$, when considered separately, are resp. in $x \ge -1$ and $x \ge 1$ (but, considering the whole function $f(z) = \sqrt{z^2-1}$, the discontinuities cancel out at $x \ge 1$ and the cut is restricted to the interval $-1 \le x \le 1$ (cf. Problem 2.16)). Then $\sqrt{z+1}$ in a point x with $-1 \le x \le 1$ takes the value $\sqrt{x+1}$ (resp. $-\sqrt{x+1}$) in the upper (resp. lower) margin of the cut, and $\sqrt{z-1}$ takes the value $i\sqrt{|x-1|}$, so that the discontinuity is $2i\sqrt{1-x^2}$ in $|x| \le 1$;
(b) $-2i\sqrt{\frac{1+x}{1-x}}$, $|x| \le 1$; *(c)* $(x-1)^\alpha\big(1 - \exp(2\pi i\alpha)\big)$, $x \ge 1$;
(d) The function $\log(z-1)$ on the upper (resp. lower) margin of the cut in $x > 1$ takes the value $\log|x-1|$ (resp. $\log|x-1| + 2\pi i$) and is continuous in all other points; the function $\log(z+1)$ on the upper (resp. lower) margin of the cut in $x < -1$ takes the value $\log|x+1| + \pi i$ (resp. $\log|x+1| - \pi i$) and is continuous in all other points. Then, the discontinuity of $f(z)$ along the cut $x > 1$ is $-2\pi i$ and along the cut $x < -1$ is $2\pi i$.
(e) $+2\pi i$; *(f)* $-2\pi i$ (See Problem 2.16)

(2.35)
The integrals along the margins of the cut produce the integral of the discontinuity: $\int_0^{+\infty} f_+(x)\,dx + \int_{+\infty}^{0} f_-(x)\,dx = \int_0^{+\infty}(f_+ - f_-)\,dx$. For the first integral:

$$\oint = 2\int_0^{+\infty} = 2\pi i\big(\mathrm{R}(i) + \mathrm{R}(-i)\big) = \pi\big(\exp(\pi i/4) - \exp(3\pi i/4)\big) = \sqrt{2}\,\pi$$

The result is $\pi/\sqrt{2}$. The two other integrals are equal to $\pi/\sqrt{3}$ and $\pi(1-i)/2\sqrt{2}$

(2.36)
In the first integral $\oint = 2\pi i\,\mathrm{R}(-1) = 2\pi i\big(-a\,\exp(i\pi a)\big)$ and the discontinuity is $\frac{x^a}{1+x^2}\big(1 - \exp(2\pi i a)\big)$; the result is $a\pi/\sin a\pi$.

The second integral is equal to $\dfrac{\pi/3}{\sin\big(\frac{\pi}{3}(b+1)\big)}$

(2.37)
(1) $2\pi/3\sqrt{3}$; $2\pi/3\sqrt{3}$
(2) The discontinuity of $\log^2 z$ is $\log x^2 - (\log x - 2\pi i)^2 = 4\pi i \log x + 4\pi^2$, then …. The given integrals are equal to 0 and $\pi/2$

(2.38)
For the first integral: if $x < 0$ then $\log x = \log|x| + i\pi$;

$$\oint \frac{\log z}{1+z^2}\,dz = \int_{-\infty}^{0} + \int_0^{+\infty} = 2\int_0^{+\infty} \frac{\log x}{1+x^2}\,dx + i\pi\int_0^{+\infty}\frac{1}{1+x^2}\,dx =$$

$$2\int \ldots + \frac{i\pi^2}{2} = 2\pi i\,\mathrm{R}(i) = \frac{i\pi^2}{2}$$

then $\int \ldots = 0$

The second integral = $-\pi^2/8\sqrt{2}$

(2.39)
The first integral is equal to $\pi/\sin a\pi$ (there is a simple pole at $z = \pi i$).

For the second integral: $\oint = 2\pi i\,\mathrm{R}(-1) = 2\pi i\big(\exp(-i\pi b)\big)$ and the discontinuity is $\frac{1}{x^b(1+x)}\big(1 - \exp(-2\pi i b)\big)$; the result is $\pi/\sin b\pi$

(2.40)
$\pi^3/8$

(2.41)
For the first integral: $\oint f(z)\,dz = 2i\int_{-1}^{+1}\ldots = 2\pi i\,\mathrm{R}_f(\infty) = \pi i$, indeed the expansion "in the neighborhood of ∞" is $\sqrt{z^2-1} = z - (1/2z) + \ldots$, then $\mathrm{R}_f(\infty) = 1/2$. The result is $\pi/2$.

The other integrals are equal to π; $\pi(2-\sqrt{3})$; $11\pi/8$

(2.42)

$$\oint f(z)\,dz = \int_{-\infty}^{+\infty} + \int_{a}^{+\infty} (f_L - f_R) i\,dy + \int_{\gamma} = 2\pi i\,\mathrm{R}_f(i) = \pi\ \log(a^2-1)$$

where f_L, f_R are the values of the integrand at the left (resp. right) margins of the cut. One has $\int_\gamma \to 0$, and $\int_a^{+\infty}(f_L - f_R)i\,dy = 2\pi i \int_a^{+\infty} 1/(y^2-1)i\,dy = \ldots$. The result is $2\pi\log(a+1)$

(2.43)
The lines $u(x, y) =$ const. are resp. hyperbolas, circumferences, parabolas; for the third case, in polar coordinates r, φ: $\mathrm{Re}\sqrt{z} = \sqrt{r}\cos(\varphi/2) = c'$, then $r(1 + \cos\varphi) = r + x = c$ which gives the family of parabolas $(c-x)^2 = x^2 + y^2$

(2.44)
(1) $u'(x', y') \propto y'$ where $y' = -i\,\mathrm{Re}\,(z') = -i\,\mathrm{Re}\,(z^2)$, then ...
(2) $u'(x', y') = (u_0/\alpha)y' = (u_0/\alpha)\mathrm{Re}\,(-iz')$, then ...

(2.45)
(3) $\widetilde{u}(r', \varphi') = \dfrac{1}{2}\mathrm{Re}\,(1+z') = \mathrm{Re}\,\dfrac{i}{i+z}$, then $u(x, y) = \dfrac{1+y}{x^2+(1+y)^2}$
(4) $\widetilde{u}(r', \varphi') = \dfrac{1}{8}\mathrm{Re}\,\big(1 - r'^2\exp(2i\varphi')\big) = \dfrac{1}{8}\mathrm{Re}\,\Big(1 - \big(\dfrac{i-z}{i+z}\big)^2\Big) = \ldots$

(2.46)
With $z' = \dfrac{-x}{x+2i}$ one gets $\left|z' + \dfrac{1}{2}\right| = \dfrac{1}{2}\dfrac{|x-2i|}{|x+2i|} = \dfrac{1}{2}$. The image of the strip is the region D in Fig. 2.8 (b)

(2.47)
The region D is transformed into the strip $0 \le y' \le 1$ where the solution is $\widetilde{u}(x', y') = y' = \mathrm{Re}\,(-iz')$; then, $u(x, y) = \mathrm{Re}\big((1-z)/(1+z)\big) = \ldots$

(2.48)
(2) Notice that the map $\Psi(z')$ diverges at $z' = -1$, i.e. $r' = 1$, $\varphi' = \pm\pi$, which corresponds to $z = \infty$. As expected, the solution $\widetilde{u}(r', \varphi')$ is a harmonic function in the interior of the circle $r' < 1$ and vanishes on the circumference $r' = 1$ (which corresponds to the boundary $y = 0$) *apart from the singularity at* $\varphi' = \pm\pi$. So, the uniqueness of solutions of the Dirichlet Problem can be recovered excluding solutions as iz, iz^2, etc. by the introduction of a boundedness condition at $z = \infty$ (the so-called "normal conditions" at the infinity)

Problems of Chap. 3

(3.1)
$\widehat{f}(\omega) = 2\frac{\sin(\omega-\omega_0)t_0}{\omega-\omega_0}$, which reaches its max value $2t_0$ at $\omega = \omega_0$

(3.2)
$\widehat{f_1}(\omega) = \pi\, a \exp(-a|\omega - \omega_0|)\,; \quad \widehat{f_2}(\omega) = \sqrt{\pi}\, a\, \exp\big(-(\omega - \omega_0)^2 a^2/4\big)$

(3.3)
Use Jordan lemma. For instance,

$$\mathscr{F}\Big(\frac{t}{(1+t^2)^2}\Big) = \int_{-\infty}^{+\infty} \frac{t\exp(i\omega t)}{(1+t^2)^2} = \begin{cases} (\text{for } \omega > 0)\ 2\pi i\, \mathrm{R}_f(i) = \frac{\pi i}{2}\omega \exp(-\omega) \\ (\text{for } \omega < 0)\ -2\pi i\, \mathrm{R}_f(-i) = \frac{\pi i}{2}\omega \exp(+\omega) \end{cases}$$

where $f = f(z) = z\exp(i\omega z)/(1+z^2)^2$.

To evaluate $\mathscr{F}^{-1}\Big(\frac{\cos\omega}{(\omega+i)^4}\Big)$, it can be convenient to use the translation theorem: $\mathscr{F}(t-a) = \exp(i\omega a)\mathscr{F}\big(f(t)\big)$, and obtain first $\mathscr{F}^{-1}\big(1/(\omega+i)^4\big)$:

$$\frac{1}{2\pi}\int_{-\infty}^{+\infty} \frac{\exp(-i\omega t)}{(\omega+i)^4} = \begin{cases} (\text{for } t < 0)\ \ 0 \\ (\text{for } t > 0)\ -2\pi i\,(1/2\pi)\,\mathrm{R}_f(-i) \end{cases} = \frac{1}{6}t^3 \exp(-t)\theta(t)$$

then

$$\mathscr{F}^{-1}\Big(\frac{\cos\omega}{(\omega+i)^4}\Big) = \frac{1}{12}\big((t-1)^3\exp(-t+1)\theta(t-1) + (t+1)^3\exp(-t-1)\theta(t+1)\big)$$

(3.4)
(1) $G(t) = \theta(t)\exp(-t)$ (2) $G(t) = -\theta(-t)\exp(+t)$
(1–2) The solutions $x_0(t) = A\exp(\mp t),\ t \in \mathbf{R}$, of the homogeneous equations $\dot{x}_0 \pm x_0 = 0$ do not admit Fourier transform if $A \neq 0$, not even as distributions in $\mathscr{S}'$, see Problem 3.69. The Green function obtained in (2) does not respect the "causality principle"; the causal $G(t)$ for this case is actually $G(t) = \theta(t)\exp(+t) = -\theta(-t)\exp(+t) + \exp(+t)$, obtained adding the solution $\exp(+t)$ (which does not admit Fourier transform).
(3) $\widehat{G}(\omega) = 1/(R - i\omega L)$; the difference in the sign depends on the initial choice in the definition of the Fourier transform. Clearly there are no consequences if one uses consistently the rules for the Fourier transform and for its inverse.

(3.5)
For instance, in the first case: $\widehat{x}(\omega) = \frac{1}{(1-i\omega)(c-i\omega)}$ and

$$x(t) = -\theta(t)\,\frac{1}{2\pi}2\pi i\big(\mathrm{R}(-ic) + \mathrm{R}(-i)\big) = \theta(t)\frac{\exp(-t) - \exp(-ct)}{c-1}$$

(3.6)
(1) $\widehat{G}(\omega) = \frac{-1/a}{(\omega-\omega_1)(\omega-\omega_2)}$ where $\omega_{1,2} = \frac{-i\,b\pm\sqrt{-b^2+4ac}}{2a}$
The Green functions can be easily found by means of Jordan lemma.
If, for instance, $4ac > b^2$ (i.e., small damping), the Green function shows oscillatory damped behavior:
$G(t) = (1/\nu)\exp(-b\,t/2a)\sin(\nu t/2a)\,\theta(t)$, where $\nu = \sqrt{4ac - b^2}$
(2) $G(t) = 2\exp(t)\sin t\,\theta(-t)$

(3.7)
(1) Only $\widehat{G}(\omega) = \exp(i\omega)/(1 - i\omega)$ gives the causal $G(t) = \theta(t-1)\exp(-(t-1))$.
(2) There are poles in both half planes $\mathrm{Im}\,z > 0$ and $\mathrm{Im}\,z < 0$; the inverse Fourier transform is a combination of terms as $\exp(z_i t)$ with support on all $t \in \mathbf{R}$, therefore $G(t)$ is not causal
(3) $\widehat{G}(\omega)$ real implies that $G(t) = \mathscr{F}^{-1}\big(\widehat{G}(\omega)\big)$ is *even*: $G(t) = G(-t)$, then $G(t)$ is not causal, apart from the case of a $G(t)$ - not belonging to $L^1(\mathbf{R})$ or $L^2(\mathbf{R})$ - having support in the point $t = 0$, e.g., $G(t) = \delta(t)$, etc.
In the case $\widehat{G}(\omega) = \exp(i\omega)\widehat{G}_1(\omega)$, a particular example where $G(t)$ is causal is $\widehat{G}(\omega) = \exp(i\omega)\frac{\sin\omega}{\omega} = \frac{\exp(2i\omega)-1}{2i\omega}$ which is the inverse Fourier transform of $G(t) = (1/2)$ for $0 < t < 2$ and 0 elsewhere.

(3.8)
(1) $\widehat{f}(\omega) \in C^0 \cap L^2(\mathbf{R})$; the functions satisfying $\widehat{f}(0) = 0$ and $\widehat{f}(q) = \widehat{f}(-q)$ are dense in $L^2(\mathbf{R})$
(2–3) The hypothesis given in (2) for $f(x)$ implies that $\widehat{f}(\omega) \in C^\infty \cap L^2(\mathbf{R})$, then ...; the further hypothesis given in (3) implies that all the derivatives $\widehat{f}^{(n)}(\omega)$ of $f(x)$ vanish at $\omega = 0$, then ...(see Problems 1.3, q.(2) and 1.6)

(3.9)
(1) $\widehat{f}(\omega) = \pi$ for $|\omega| < 1$ and $= 0$ elsewhere, then $(f, f) = \frac{(\widehat{f},\widehat{f})}{(2\pi)} = (2\pi^2)/(2\pi) = \pi$
(2)(*a*) $\widehat{f}_n(\omega) \propto \omega^n$ for $|\omega| < 1$ and $= 0$ elsewhere, which are a complete set in $L^2(-1, 1)$ but not in $L^2(\mathbf{R})$, therefore any $h(x)$ such that the support of its Fourier transform $\widehat{h}(\omega)$ has no intersection with the interval $|\omega| < 1$ satisfies $(h, f_n) = 0$
(*b*) $\|f_n(x)\|^2 = \|\omega^n \widehat{f}(\omega)\|^2/(2\pi) = \pi^2/(2\pi)\int_{-1}^{+1}\omega^{2n}d\omega = \pi/(2n+1)$
(3)(*a*) $g_n(x) = (-1)^n\frac{\sin(x-n\pi)}{x-n\pi}$, then the functions $\widehat{g}_n(\omega) = (-1)^n\pi\exp(in\pi\omega)$ for $|\omega| < 1$ and $= 0$ elsewhere, are orthogonal in $L^2(\mathbf{R})$ and complete in $L^2(-1, 1)$ but not in $L^2(\mathbf{R})$, see before for the functions $h(x)$
(*b*) See Problem 3.9

(3.10)
(1) $I(a) = 2\pi\int_0^a \exp(-t)\,dt \to 2\pi$
(2) $I(a) = 2\pi\int_0^a \exp(-t^2/4)\,dt\frac{1}{2\sqrt{\pi}} \to \pi$

(3.11)
(2) $\widehat{f}(-\omega) = \mathscr{F}\big(f(-t)\big)$

(3–4) $I = 0$, immediate (Jordan lemma) and confirmed observing that the inverse Fourier transforms $f_1(t)$ and $f_2(t)$ have disjoint supports

(3.12)
(1) $v(t) = \theta(t)\dfrac{\exp(-\beta t) - \exp(-t)}{1-\beta}$
(2) $W_f = W_\beta = 1/(2+2\beta)$
(3) E.g., $(f, v) = \dfrac{1}{2\pi}(\widehat{f}, \widehat{v}) = \dfrac{1}{2\pi}\displaystyle\int_{-\infty}^{+\infty} \dfrac{d\omega}{(1+i\omega)(1-i\omega)(\beta - i\omega)} = i\,\mathrm{R}(i) = 1/(2+2\beta)$

(3.13)
(2) $(f, v) = \dfrac{1}{2\pi}\displaystyle\int_{-\infty}^{+\infty} (\beta + i\omega)\widehat{v}^*(\omega)\widehat{v}(\omega)\,d\omega = \dfrac{\beta}{2\pi}\displaystyle\int_{-\infty}^{+\infty} \widehat{v}(-\omega)\widehat{v}(\omega)\,d\omega = \beta(v, v)$
observing that $\omega\widehat{v}^*(\omega)\widehat{v}(\omega) = \omega\widehat{v}(-\omega)\widehat{v}(\omega)$ is an odd function

(3.14)
(1) $\|\widehat{b}(\omega)\|^2_{L^2(\mathbf{R})} \le \sup_{\omega\in\mathbf{R}} |\widehat{G}(\omega)|^2 \|\widehat{a}(\omega)\|^2_{L^2(\mathbf{R})}$, then $C = \sup_{\omega\in\mathbf{R}} |\widehat{G}(\omega)| \le \|G(t)\|_{L^1(\mathbf{R})}$
(2) $\widehat{a}(\omega)$ and $\widehat{G}(\omega) \in L^2(\mathbf{R})$, the product of two functions $\in L^2(\mathbf{R})$ belongs to $L^1(\mathbf{R})$, then $b(t) = \mathscr{F}^{-1}\big(\widehat{G}(\omega)\widehat{a}(\omega)\big)\ \ldots$
(3) $\widehat{b}(\omega)$ rapidly goes to zero as $|\omega| \to \infty$, then $b(t) \in C^\infty$. But one can conjecture that $b(t)$ has no compact support, this indeed would imply $\widehat{b}(\omega) \in C^\infty$, which is not the case in general. Actually, it can be shown that the Fourier transform of a function with compact support cannot have compact support; for a proof, see Problem 3.121 q.(2)

(3.15)
(2) $\Delta t = \frac{1}{\sqrt{2}\alpha}$, $\Delta\omega = \frac{\alpha}{\sqrt{2}}$

(3.16)
(1) Use the identities

$$\frac{\|\widehat{T}\,\widehat{f}\|}{\|\widehat{f}\|} = \frac{\|\widehat{g}\|}{\|\widehat{f}\|} = \frac{2\pi\,\|g\|}{2\pi\,\|f\|} = \frac{\|T\,f\|}{\|f\|}$$

(3) $\psi = \mathscr{F}^{-1}(\varphi)$ is eigenvector of T with eigenvalue λ. See also Problem 1.43, q.(5)
(4) Yes: See Problem 1.43, q.(2)

(3.17)
For instance, with $T = d/dx$, i.e., $\mathscr{F}\big(df(x)/dx\big) = -i\omega\widehat{f}(\omega)$, then $\widehat{T} = -i\omega$ (notice that both T and $\widehat{T}$ are unbounded)

(3.18)

$$\widehat{T}\,\widehat{f}(\omega) = \begin{cases} \widehat{f}(\omega) & \text{for } |\omega| < 1 \\ 0 & \text{for } |\omega| > 1 \end{cases}$$

(1) The subspace of the functions $f(t)$ such that the support of their Fourier transform $\widehat{f}(\omega)$ is contained in the interval $|\omega| \le 1$
(2) T is the ideal "filter" for "low frequencies", $|\omega| \le 1$
(3) $\widehat{g}(\omega) \in L^1(\mathbf{R})$ and vanishes rapidly as $|\omega| \to \infty$, then $g(t) \in C^\infty,\ g(t) \to 0$ as $|t| \to \infty$
(4) Strong convergence to the identity operator: $\|(I - \widehat{T}_n)\widehat{f}\|^2 = \int_{|\omega|>n} |\widehat{f}(\omega)|^2\, d\omega \to 0$, but $\|I - \widehat{T}_N\| = 1$

(3.19) $$\widehat{T}\,\widehat{f}(\omega) = \pi\ \exp(-|\omega|)\,\widehat{f}(\omega)$$
(1) $\|T\| = \|\widehat{T}\| = \sup_\omega \pi\ \exp(-|\omega|) = \pi$
(2) $\mathrm{Ran}\,T$ is the set of the functions $g(x)$ such that $\widehat{g}(\omega)\exp(+|\omega|) \in L^2(\mathbf{R})$, then $\mathrm{Ran}\,T \ne L^2(\mathbf{R})$, but dense in it: $\overline{\mathrm{Ran}\,T} = L^2(\mathbf{R})$
(3) $(\widehat{T} - \rho I)\widehat{f}(\omega) = \big(\pi\ \exp(-|\omega|) - \rho\big)\widehat{f}(\omega)$, then $\rho \notin [0, \pi]$
(4) $g(x) \in C^\infty$
(5) Yes, indeed $\widehat{g}_n(\omega) = \pi\ \exp(-|\omega|)\,\widehat{f}_n(\omega)$ and $\exp(-|\omega|) \ne 0$; see Problem 3.16, q.(4)
(6) $\|(\widehat{T}_a - \pi I)\widehat{f}(\omega)\|^2 = \pi^2 \int_{-\infty}^{+\infty} |\widehat{f}(\omega)|^2 \big(\exp(-a|\omega|) - 1\big)^2 d\omega \to 0$,
thanks to Lebesgue theorem, indeed $|\widehat{f}(\omega)|^2 \in L^1(\mathbf{R})$, etc., then strong convergence to πI; whereas
$$\|T_a - \pi I\| = \|\widehat{T}_a - \pi I\| = \pi\ \sup_\omega \big(\exp(-a|\omega|) - 1\big) = \pi$$

(3.20)
(1) $$\big(g, T_a f\big) \propto \int_{-\infty}^{+\infty} \exp(ia\omega)\widehat{g}^*(\omega)\widehat{f}(\omega)\, d\omega \to 0$$
thanks to Riemann-Lebesgue lemma for the Fourier transform in $L^1(\mathbf{R})$ $\big($recall that the product of two functions $\in L^2(\mathbf{R})$ belongs to $L^1(\mathbf{R})\big)$
(2) Use Lebesgue theorem, see previous problem, q.(6)

(3.21)
(1) It is convenient to use two completely "neutral" variables $\xi,\ \eta$:
$$\widetilde{\mathscr{F}}^2\big(f(\xi)\big) = \widetilde{\mathscr{F}}\big(\widehat{f}(\eta)\big) = \frac{1}{\sqrt{2\pi}} \int_{-\infty}^{+\infty} \widehat{f}(\eta)\exp(i\xi\eta)\, d\eta =$$
$$\frac{1}{\sqrt{2\pi}} \int_{-\infty}^{+\infty} \widehat{f}(\eta)\exp\big(-i(-\xi)\eta\big)\, d\eta = f(-\xi)$$

(2) For instance, $\mathscr{F}\big(d^2 f(x)/dx^2\big) = -\omega^2 \widehat{f}(\omega)$, etc.
(3–4) $\widetilde{\mathscr{F}}$ has eigenvalues $+1, i, -1, -i$ with eigenfunctions resp. the Hermite functions $u_{4m}(x),\ u_{4m+1}(x),\ u_{4m+2}(x),\ u_{4m+3}(x)\,;\ m = 0, 1, 2, \ldots$.

(3.22)
(1) The pointwise limit is zero, *for all* $x \in \mathbf{R}$. But $< f_n, \varphi >=$
$\int_0^{\pi/n} \sin nx\, \varphi(x)\, dx = \int_0^\pi \sin t\, \varphi(t/n)\, dt \to 2\varphi(0)$, then the $\mathscr{S}'$-limit is $2\delta(x)$

(3.23)
(1) *(b)* The $\mathscr{S}'$-limit of the Fourier transforms is the constant function $= 1$; indeed $\widehat{f_n}(y) = \frac{n}{n-iy}$ tends pointwise to 1, and $|\widehat{f_n}(y)| \le 1$, then also $< \widehat{f_n}, \varphi > \to < 1, \varphi >, \forall \varphi \in \mathscr{S}$. Therefore, the $\mathscr{S}'$-limit of the given sequence is $\delta(x)$
(2) $< f_n^{(1)}, \varphi > = < n \exp(-n|x|), \varphi(x) > = \int_{-\infty}^{+\infty} \ldots = \int_{-\infty}^{+\infty} \exp(-|t|)\varphi(t/n)\, dt \to 2\,\varphi(0)$, then $f_n^{(1)}(x) \to 2\,\delta(x)$.
Using Fourier transform: $\mathscr{F}\big(f_n^{(1)}(x)\big) = \frac{2n^2}{n^2+y^2} \to 2,\ f_n^{(1)}(x) \to 2\,\delta(x)$;
$\mathscr{F}\big(f_n^{(2)}(x)\big) = \sqrt{\pi}\exp(-y^2/(4n^2)) \to \sqrt{\pi},\ f_n^{(2)}(x) \to \sqrt{\pi}\,\delta(x)$;
$\widehat{f_n^{(3)}}(y) \to \pi,\ f_n^{(3)}(x) \to \pi\,\delta(x)$
(3) $g_n(x) \to -(\sqrt{\pi}/2)\,\delta'(x)$
(4) See next question
(5) *(a)* $< u_n, \varphi > = n \int_0^{1/n} \varphi(x)\, dx - \int_n^{n+1} \varphi(x)\, dx = \varphi(x_1) - \varphi(x_2)$ with $0 < x_1 < 1/n$ and $n < x_2 < n+1$ thanks to the mean value problem; then, at $n \to \infty$, this quantity $\to \varphi(0)$, or $u_n(x) \to \delta(x)$.
Using Fourier transform: $\widehat{u}_n(y) = \frac{\exp(iy/n)-1}{iy/n} - \exp(iyn)\frac{\exp(iy)-1}{iy} \to 1$
(b) Nothing changes, indeed $\pm n^\alpha \varphi(x_2) \to 0$ because $\varphi(x)$ is rapidly vanishing at $x \to \infty$. The same for the Fourier transform, indeed $\pm n^\alpha \exp(iyn) \to 0$ in $\mathscr{S}'$, because $< \pm n^\alpha \exp(iyn), \varphi > = < \exp(iyn), \pm n^\alpha \varphi >$ can be seen as a Fourier transform (recall that both φ and $\widehat{\varphi}$ are rapidly vanishing)

(3.24)
(1) $\widehat{f_n}(y) \to \pi$ pointwise and $|\widehat{f_n}(y)| \le$ const, then also $< \widehat{f_n}, \varphi > \to < \pi, \varphi >$ and $f_n(x) \to \pi\,\delta(x)$
(2) $< \mathscr{F}(\delta), \varphi > = < \delta, \widehat{\varphi} > = \widehat{\varphi}(0) = \int_{-\infty}^{+\infty} \varphi(x)\, dx = < 1, \varphi >$. This is one of the few cases where this formal definition of $\mathscr{F}T$ can be used in practice for finding explicitly the Fourier transform of a distribution

(3.25)
(1) $f_{t_0}(t) \to \exp(-i\omega_0 t)$; $\widehat{f_{t_0}}(\omega) = 2\dfrac{\sin(\omega-\omega_0)t_0}{\omega-\omega_0} \to 2\pi\,\delta(\omega-\omega_0)$ as $t_0 \to \infty$
(see previous problem with n replaced by t_0), i.e., the frequency contribution is "concentrated" in $\omega = \omega_0$
(2) For the wave $f_2(t)$ in Problem 3.2, one has
$\widehat{f_2}(\omega) = \sqrt{\pi}\, a \exp\big(-(\omega-\omega_0)^2 a^2/4\big) \to 2\pi\,\delta(\omega-\omega_0)$, see Problem 3.23, q.(2) with x replaced by $\omega - \omega_0$ and n replaced by $a/2$

(3.26)
(1) $\widehat{u}(\omega) = \mathscr{F}\big(\theta(t) - \theta(t-1)\big) = i\,\mathrm{P}(1/\omega) + \pi\delta(\omega) - \big(i\,\mathrm{P}(1/\omega) + \pi\delta(\omega)\big) \exp(i\omega) = i\,\mathrm{P}\frac{1}{\omega}\big(1 - \exp(i\omega)\big) + \pi\delta(\omega)\big(1 - \exp(i\omega)\big) = \frac{\exp(i\omega)-1}{i\omega}$
(2) $\widehat{v}(\omega) = -i\mathrm{P}(1/\omega) + \pi\delta(\omega) + \exp(i\omega)\big(i\mathrm{P}(1/\omega) + \pi\,\delta(\omega)\big) = 2\pi\,\delta(\omega) - \frac{\exp(i\omega)-1}{i\omega}$

(3.27)
The first derivative of $\widehat{f}(k) = \mathscr{F}\big(f(x)\big)$ is $\pi\ \mathrm{sgn}(k)\exp(-|k|)$, which is discontinuous at $k = 0$, then.... $\mathscr{F}\big(g(x)\big) = 2\pi\delta(k) - \pi\ \exp(-|k|)$

(3.28)
Some examples:

$\mathscr{F}\big(|t|\big) = \mathscr{F}(t\,\mathrm{sgn}\,t) = 2\,D\,\mathrm{P}\frac{1}{\omega}$;

$\mathscr{F}^{-1}\Big(\mathrm{P}\frac{\omega}{\omega-1}\Big) = i\,D\mathscr{F}^{-1}\Big(\mathrm{P}\frac{1}{\omega-1}\Big) = \frac{1}{2}D\big(\exp(-it)\mathrm{sgn}\,t\big) = -\frac{i}{2}\exp(-it)\mathrm{sgn}\,t + \delta(t)$;

$\mathscr{F}\big(\exp(i\,|t|)\big) = \mathscr{F}\big(\theta(t)\,\exp(it)\big) + \ldots = -2i\,\mathrm{P}\Big(\frac{1}{\omega^2-1}\Big) + \pi\big(\delta(\omega+1) + \delta(\omega-1)\big)$;

$\mathscr{F}^{-1}\Big(\mathrm{P}\Big(\frac{1}{\omega}\Big)\frac{1}{\omega+i}\Big) = -\frac{1}{2}\mathrm{sgn}\,t + \theta(t)\exp(-t)$, which can be obtained either using Jordan lemma (introducing an "indentation" at $\omega = 0$), or writing $\mathrm{P}\Big(\frac{1}{\omega}\Big)\frac{1}{\omega+i} = -i\,\mathrm{P}\frac{1}{\omega} + \frac{i}{\omega+i}$

(3.29)
(1) $\mathscr{F}\big(g(x)\big) = \pi\big(2\delta(k) - \delta(k-1) - \delta(k+1)\big)$

$$\widehat{f}(k) = \begin{cases} -\pi i & \text{for } -1 < k < 0 \\ \pi i & \text{for } 0 < k < 1 \\ 0 & \text{elsewhere} \end{cases}$$

(2) $$\widehat{F}(k) = \begin{cases} \pi(k+1) \text{ for } -1 \le k \le 0 \\ \pi(-k+1) \text{ for } 0 \le k \le 1 \\ 0 \quad \text{elsewhere} \end{cases}$$

$\widehat{F}(k)$ is continuous because $F(x) \in L^1(\mathbf{R})$

(3.30)
The Fourier transform is

$$(\pi/2)\exp(i\omega\pi)\big(\exp(ia)\mathrm{sgn}(\omega+1) - \exp(-ia)\mathrm{sgn}(\omega-1)\big)$$

If $a = n\pi$, $n \in \mathbf{Z}$, the symbol P can be omitted, the distribution becomes a function $\in L^2(\mathbf{R})$ and the support of its Fourier trasform is $|\omega| \le 1$

(3.31)
(1) Start with (for instance) $\mathscr{F}\big(-i\,\theta(t)\exp(-\varepsilon t)\big) = \frac{1}{\omega+i\,\varepsilon}$; then for $\varepsilon \to 0^+$ one gets $\mathscr{F}\big(-i\,\theta(t)\big) = \mathrm{P}\frac{1}{\omega} - i\pi\,\delta(\omega)$. The result is reached replacing ω with a "neutral" variable x
(2) $\mp i\pi$

(3.32)
(1) $\mathscr{F}^{-1}\big(u_a(\omega)\big) = 1$ for $0 < \omega < a$ and $= 0$ elsewhere,

$\lim_{a\to\infty} u_a(\omega) = \mathscr{F}\big(\theta(t)\big) = i\,\mathrm{P}(1/\omega) + \pi\delta(\omega)$

(2) π. Notice that this limit cannot be obtained by means of usual Lebesgue theorem or by integration in the complex plane
(3) $i\pi$

(3.33)
(2) $\widehat{u}_a(y) = 2i\exp(iay)\mathrm{P}(1/y) \to -2\pi\delta(y)$
(3) $i\pi$
(4) $-\pi/2$
(5) $I_a = -\pi/2 + \pi\exp(-2a)$

(3.34)
(2) $\lim_{\varepsilon\to 0^+} u_\varepsilon(x) = -D\,\mathrm{P}\frac{1}{x} + i\,\pi\delta'(x)$
(3) $\lim\int \ldots =< -D\,\mathrm{P}\frac{1}{x} + i\,\pi\delta'(x), \exp(-x^2) >=< \mathrm{P}\frac{1}{x}, (-2x)\exp(-x^2) >= -2\sqrt{\pi}$; and resp. $i\pi$

(3.35)
(1–2) $\widehat{T}_a = (\pi\, i/a)\big(\exp(i\,a\,\omega) - 1\big)\mathrm{sgn}\omega \to -\pi\,|\omega| = \widehat{T}$, and $T = -D\,\mathrm{P}(1/x)$
(3) $-2\sqrt{\pi}$ and resp. 0

(3.36)
(*b*) $< x^2 D\,\mathrm{P}\frac{1}{x}, \varphi >=< D\,\mathrm{P}\frac{1}{x}, x^2\varphi >=< \mathrm{P}\frac{1}{x}, -2x\,\varphi - x^2\varphi' >=$
$-2\int_{-\infty}^{+\infty}\varphi(x)\,dx - \int_{-\infty}^{+\infty} x\,\varphi'(x)\,dx = -\int_{-\infty}^{+\infty}\varphi(x)\,dx =< -1, \varphi >$.
From Fourier transform: $\frac{d^2}{dy^2}|y| = 2\,\delta(y)$
(*c*) Use the same procedure; notice that $\int_{-\infty}^{+\infty} x\,\varphi''(x)\,dx = -\int_{-\infty}^{+\infty}\varphi'(x)\,dx = 0$, for the properties of the functions $\varphi(x) \in \mathscr{S}$.
The Fourier transform gives $\frac{d^2}{dy^2}(y|y|) = 2\,\mathrm{sgn}\,(y)$

(3.37)
One has $\mathscr{F}(C(f)) = 2i\frac{\omega}{1+\omega^2}\widehat{f}(\omega)$, then:
(*a*) $2i\,\widehat{f}(\omega) = (1+\omega^2)\Big(\mathrm{P}\frac{1}{\omega} + c\,\delta(\omega)\Big)$
(recall that $y\,T = 1$ is solved by $T = \mathrm{P}\frac{1}{y} + c\,\delta(y)$, where $c =$ arbitrary const.), then $f(x) = -(1/4)\mathrm{sgn}\,x + (1/2)\delta'(x) +$ const;
(*b*) $\big(\frac{3}{4} - x^2\big)\exp(-x^2) +$ const;
(*c*) $\frac{\omega^2-4\omega+1}{2(1+\omega^2)}\widehat{f}(\omega) = 0$
(recall that $(y-a)\,T = 0$ is solved by $T = c\,\delta(y-a)$), then
$f(x) = c_1\cos(\omega_1 x) + c_2\sin(\omega_2 x)$ where $c_1,\ c_2 =$const and $\omega_{1,2} = 2 \pm \sqrt{3}$;
(*d*) $f(x) = 0$

(3.38)
(1) $g_a(x) \to f(x)$ in the $L^2(\mathbf{R})$ norm
(2) $g_a(x) \to \mathrm{P}(1/x)$ in $\mathscr{S}'$
(3) $g_a(x) \to \delta'(x)$ in $\mathscr{S}'$

(3.39)
(1) $\pi i\big(\exp(iax) - 1\big)/x$
(2) $|F_a(x)| \le \|\widehat{F}_a(\omega)\|_{L^1(\mathbf{R})} = a\,;\ F_a(x) \in L^2(\mathbf{R}),\ \notin L^1(\mathbf{R})$
(3) (*b*) $F_a(x) = (\cos ax - 1)/x$
(4) $(1 - \cos ax)/x \to \mathrm{P}(1/x)$

(3.40)
$\lim_{\varepsilon\to 0^+} f_\varepsilon(t) \;=\; \theta(t) + \exp(t)\theta(-t)$

(3.41)
(1) $f_\varepsilon^{(+)}(t) = (-i)\big(\mathrm{R}\,(1-i\varepsilon) + \mathrm{R}\,(-1-i\varepsilon)\big) = \exp(-\varepsilon t)\sin t\,\theta(t) \to F^{(+)}(t) = \theta(t)\sin t$ and $F^{(-)}(t) = -\theta(-t)\sin t$
(2) E.g., $G^{(+)}(\omega) = \frac{1}{2}\mathrm{P}\Big(\frac{1}{\omega+1}\Big) - \frac{1}{2}\mathrm{P}\Big(\frac{1}{\omega-1}\Big) - \pi i\,\delta(\omega+1) + \pi i\,\delta(\omega-1)$

(3.42)
(1) $f_\varepsilon^{(+)}(t) = \frac{1}{2(i\varepsilon-1)}\exp\big((i-\varepsilon)|t|\big) \to F^{(+)}(t) = -(i/2)\exp(i|t|)$ and $F^{(-)}(t) = (i/2)\exp(-i|t|)$

The four functions $F^{(\pm)}(t)$ obtained in this and in the above problem are all different, although putting "naively" $\varepsilon = 0$ in $g_\varepsilon^{(\pm)}(\omega)$ one would obtain in all cases $1/(1-\omega^2)$ (which is *not* a distribution! see next question)
(3) $H(t) = (1/2)\sin|t| = (1/2)\sin t\ \mathrm{sgn}\, t$, which is different from all the functions obtained in these two problems

(3.43)
(1–3) $\widehat{F}_n(\omega) = \pi\delta(\omega) + i\mathrm{P}(1/\omega) - 1/(n-i\omega) \to \mathscr{F}\big(\theta(x)\big)$, etc.

(3.44)
(1) $\lim = -i\omega$, in $\mathscr{S}'$
(2) $f_a(x) = (1/a^2)\begin{cases} 1 & \text{for } -a < x < 0 \\ -1 & \text{for } 0 < x < a \\ 0 & \text{for } |x| > a \end{cases}$

The symbol P can be omitted because $\widehat{f_a}(\omega)$ is continuous in $\omega = 0$
(3) $\lim = \delta'(x)$
(4) $< f_a, \varphi >= \big(\exp(-a^2) - 1\big)/a^2 \to -1 =< \delta', \varphi >$

(3.45)
(1) $\widehat{F}_L(\omega) \;=\; 2i\,\mathrm{P}\frac{\sin L\omega}{\omega^2}$
(3) $-2\pi i\delta'(\omega)$

(3.46)
(1) $\widehat{G}(\omega) = ig(\omega)\mathrm{P}(1/\omega) + \text{const} \times \delta(\omega)$
(2) $f(x){=}F'(x){=}u(x) - \delta(x-1)$; $g(\omega){=}\mathscr{F}\big(f(x)\big) = \dfrac{\exp(i\omega) - 1 - i\omega\exp(i\omega)}{i\omega}$

$G(\omega) = \mathscr{F}\big(F(x)\big) = \dfrac{\exp(i\omega) - 1 - i\omega\exp(i\omega)}{\omega^2}$

$G(\omega) \in C^\infty$ because $F(x)$ is rapidly (!) vanishing at $x \to \pm\infty$
(4) $\mathscr{F}\big(F_1(x)\big) = \frac{\exp(i\omega)-1}{\omega}\mathrm{P}\big(\frac{1}{\omega}\big) + \pi\delta(\omega)$

(3.47)
(1) $\mathscr{F}(\arctan x) = i\pi\exp(-|\omega|)\mathrm{P}(1/\omega)$
(2) $(\pi/2)\ \mathrm{sgn}\, x$, and ...

(3.48)
(2) $\mathscr{F}\big(\mathrm{erf}(t)\big) = i\sqrt{\pi}\exp(-\omega^2/4)\mathrm{P}(1/\omega) + \pi^{3/2}\delta(\omega)$
(3)

$$\mathscr{F}\Big(\int_{-\infty}^{t} \dots\Big) = \mathscr{F}(\arctan t + \pi/2) = \mathscr{F}(\arctan t) + \pi^2\delta(\omega)$$

$$= \mathscr{F}\big(\theta(t)\big)\,\mathscr{F}(1/1+t^2) = \big(i\mathrm{P}(1/\omega) + \pi\delta(\omega)\big)\big(\pi\exp(-|\omega|)\big) \quad \text{etc.}$$

(3.49)
(1) $F_\varepsilon(t) = \exp(-n\varepsilon)$ for $n < t < n+1\,;\ n = 0, 1, \dots$
(2) $\widehat{F}_\varepsilon(\omega) = \frac{1-\exp(i\omega)}{i\omega\left(1-\exp(-\varepsilon+i\omega)\right)}$
(3) $\| F_\varepsilon(t)\|^2 = 1/\big(1-\exp(-2\varepsilon)\big)$
(4) $\lim_{\varepsilon\to 0}\widehat{F}_\varepsilon(\omega) = i\mathrm{P}(1/\omega) + \pi\delta(\omega)$

(3.50)
(1) Using, e.g., the example $u_\varepsilon(x)$ given in Problem 3.23, q.(4), one has
$< u_\varepsilon^2(x), \varphi >= \frac{1}{4\varepsilon^2}\int_{-\varepsilon}^{+\varepsilon}\varphi(x)\,dx = \frac{1}{2\varepsilon}\varphi(\xi)$, with $-\varepsilon < \xi < \varepsilon$
(thanks to the mean value theorem), then ...

(3.51)
The convolution product is not defined for $a = \pm 1$; this is clear both from the convolution product in the variable x, with the presence of the divergent integral $\int_{-\infty}^{+\infty}(\sin^2 y)/y\,dy$, and from the Fourier transform, with the presence of a $\delta(k\pm 1)$ multiplied by a function with discontinuity in $k = \pm 1$

(3.52)

$$(1)\ v_\varepsilon(x) = \frac{2}{\varepsilon}\begin{cases} -1+\exp(\varepsilon x) & \text{for } x < 0 \\ 1-\exp(-\varepsilon x) & \text{for } x > 0 \end{cases}$$

(2) $u_\varepsilon(x) \to 1,\ v_\varepsilon(x) \to 2x$

$$(3)\ v_{\varepsilon_1,\varepsilon_2}(x) = \frac{2}{\varepsilon_1\varepsilon_2}\begin{cases} \varepsilon_2\exp(\varepsilon_1 x) - \frac{\varepsilon_1+\varepsilon_2}{2} & \text{for } x < 0 \\ -\varepsilon_1\exp(-\varepsilon_2 x) + \frac{\varepsilon_1+\varepsilon_2}{2} & \text{for } x > 0 \end{cases}$$

(3.53)
(1) $< x\,\delta'(x), \varphi(x) >=< \delta', x\,\varphi >= -< \delta, \frac{d}{dx}(x\,\varphi) >= -\varphi(0)$
The Fourier transforms: $d^2(1)/d\omega^2 = d^2(\omega)/d\omega^2 = 0;\ d^2(\omega^2)/d\omega^2 = 2$
(3) *(b)* $h(x)\delta'(x-x_0) = -h'(x_0)\delta(x-x_0) + h(x_0)\delta'(x-x_0)$
(c) $-2\pi\big(i\,\delta'(\omega) + \delta(\omega)\big) = -2\pi i\big(\exp(i\omega)\delta'(\omega)\big)$

(3.54)
(1) *(b)* $T = \mathrm{P}(1/y) + c\,\delta(y)\,$; *(d)* $T = -\frac{d}{dy}\mathrm{P}(1/y) + c_0\delta(y) + c_1\delta'(y)\,$;
(e) $T = c_0\delta(y) + c_1\delta'(y) + c_2\delta''(y)\,$; *(g)* $T = \frac{\sin y}{y} + c\delta(y)\,$;
(h) $T = \frac{1-\cos y}{y^2} + c_0\delta(y) + c_1\delta'(y)$

(2) For the equations *(g)* and *(h)* there are (nonzero) solutions $\in L^2(\mathbf{R})$, obtained choosing the arbitrary constants equal to zero

(3.55)
(b) $T = \mathrm{P}\big(1/(y-1)\big) + c\,\delta(y-1)$; *(c)* $T=0$; *(d)* $T = 1/(y \pm i)$

(3.56)
(b) $T = \left(\mathrm{P}\frac{1}{y-1} - \mathrm{P}\frac{1}{y}\right) + c_0\delta(y) + c_1\delta(y-(1)$;
(d) $T = \frac{1}{2}\left(\mathrm{P}\frac{1}{y-1} - \mathrm{P}\frac{1}{y+1}\right) + c_1\delta(y-(1) + c_2\delta(y+1)$; *(h)* $T = 1/(1+y^2)$
only for the last equation there is a (nonzero) solution $\in L^2(\mathbf{R})$

(3.57)
(a) $T = \sum_{n\in\mathbf{Z}} c_n\delta(y-n\pi)$, where c_n can be "almost arbitrary", not only $\in \ell^2$ but also e.g.. polynomially divergent;
(c)

$$T = \sum_{n\in\mathbf{Z}} c_n\delta(y-2n\pi) + \sum_{m\in\mathbf{Z}} c_m\delta'(y-2m\pi)$$

(d) $T = c_0\delta(y) + c_1\delta'(y) + c_2\delta''(y)$
(e) $T =$ any finite combination of $\delta^{(n)}(y),\ n = 0, 1, 2, \dots$

(3.58)
(a) $T = -\delta'(y) + c\,\delta(y)$ (see Problem 3.53, q.(1)); *(b)* $T = -\delta(y) + c\,\delta(y-1)$;
(c) $T = (1/2)\delta''(y) + c_0\delta(y) + c_1\delta'(y)$;
(d) $\mathcal{F}(y\,T) = -i\,D\widehat{T} = i\,\pi\,\mathrm{sgn}\,k$, then $\widehat{T} = -\pi|k| + c$; on the other hand $\mathcal{F}\big(D\,\mathrm{P}(1/y)\big) = \pi|k|$, therefore: $T = -D\,\mathrm{P}(1/y) + c\,\delta(y)$

(3.59)
(a) $\big(-i\omega + \exp(i\omega\pi/2)\big)\widehat{u}(\omega) = 0$, i.e., $\omega = \pm 1$ and $\widehat{u}(\omega) = c'\delta(\omega-1) + c''\,\delta(\omega+1)$, then $u(x) = c_1\cos x + c_2\sin x$
(b) $(-\omega^2 + 1 - 2\cos\omega\pi)\widehat{u}(\omega) = 0$, and the same solution as in *(a)*
(c) any combination of $\exp(2\pi i n x/a)$;
(d) $u = c_0 + c_1 x + c_2 x^2$

(3.60)
One obtains $\omega\,D\widehat{T} = 0$ and then $T = i\,c_1\mathrm{P}(1/x) + c_2\,\delta(x)$

(3.61)
The coefficients of the Fourier expansion of $f_\varepsilon(x)$ are $\frac{\sin n\varepsilon}{n\varepsilon}$, then …

(3.62)
(2) $f'(x) = 1 - 2\pi\sum_{m\in\mathbf{Z}}\delta(x-2m\pi) = \dots$, see Problem 1.21, q.(2)

(3.63)
(1)*(b)*

$$S_n f = \int_{-\infty}^{+\infty} f(y)\Big(\delta(x-y) - \frac{\sin\big(n(x-y)\big)}{\pi(x-y)}\Big)\,dy$$

(3.64)
(1) For any $|\lambda| \le 1$ there are infinite "eigenvectors" of the form $\delta(x - x_k)$, where $\sin x_k = \lambda$
(2) $\widehat{T}\,\widehat{f}(\omega) = \exp(i\omega)\widehat{f}(\omega) = \lambda \widehat{f}(\omega)$, etc. If $\lambda = 1$ then $\widehat{f}(\omega)$ are combinations of $\delta(\omega - 2n\pi)$, $n \in \mathbf{Z}$, and the "eigenvectors" of T are (expectedly!) the periodic functions with period 1, see also Problem 3.57 *(a)*

(3.65)
(1) $\widehat{T}\,\widehat{f}(\omega) = \big(1 - \exp(i\omega)\big)\widehat{f}(\omega)$. $\|T\| = \|\widehat{T}\| = 2$.
There is no $f_0(x)$ with the requested property (see Problem 1.78)
(2) No eigenvectors in $L^2(\mathbf{R})$, but there are "eigenvectors" in $\mathscr{S}'$ (see problem above) for all $|\lambda| \le 2$
(3) $\mathtt{Ker}\,T = \{0\}$, $\mathtt{Ran}\,T \ne L^2(\mathbf{R})$, but dense in it: $\overline{\mathtt{Ran}\,T} = L^2(\mathbf{R})$
(4) T_{ab} converges strongly to zero: put $\varepsilon = b - a$, then

$$\|\widehat{T}_{ab}\widehat{f}(\omega)\|^2 = \|\big(1 - \exp(i\omega\varepsilon)\big)\widehat{f}(\omega)\|^2 = \int_{-\infty}^{+\infty} |1 - \exp(i\omega\varepsilon)|^2 |\widehat{f}(\omega)|^2\, d\omega \to 0$$

thanks to Lebesgue theorem.
There is no norm convergence, indeed $\|\widehat{T}_{ab}\| = \sup_\omega |1 - \exp(i\omega\varepsilon)| = 2$

(3.66)
(1) $G(t) = 2\,\theta(-t)\,\exp(t)$, not causal
(2) $\|T\| = \|\widehat{T}\| = 2$, no eigenfunctions $\in L^2(\mathbf{R})$
(3) The expression $b(t) = \big(G * a\big)(t)$ ensures "time-invariance" of the system, i.e., knowing $T\big(a(t)\big) = b(t)$ one *also* knows $T\big(a(t+\tau)\big) = b(t+\tau)$ for any $\tau \in \mathbf{R}$, then it is enough that the set of the functions $a(t+\tau)$ contains a complete set in $L^2(\mathbf{R})$, ...

(3.67)
(a) $G(t)$ uniquely defined but $\in \mathscr{S}'$, $\notin L^2(\mathbf{R})$.
(b) The only information is $\widehat{G}(\omega) = 0$ for $|\omega| < 1$.
(c) $\widehat{G}(\omega) = 1$ for $|\omega| < 1$ and $\widehat{G}(\omega) = 0$ for $1 < |\omega| < 2$, undetermined for $|\omega| > 2$.
(d) Impossible.
(e) $G(t)$ is determined apart from an arbitrary additive constant; there is *one* $G(t) \in L^2(\mathbf{R})$.
(f) The same arbitrariness as in *(e)*, but there are no $G(t) \in L^2(\mathbf{R})$.
(g) $G(t)$ is uniquely defined in the hypothesis $G(t) \in L^2(\mathbf{R})$, otherwise $\widehat{G}(\omega)$ contains arbitrary combinations of delta functions, then $G(t)$ ...
(h) The same as in *(g)*, now $G(t) = 0$ is the only Green function $\in L^2(\mathbf{R})$.

(3.68)
(2) $\|T\| = \|\widehat{T}\| = \sup_{\omega\in\mathbf{R}} |\widehat{G}(\omega)| \le \|G(t)\|_{L^1(\mathbf{R})}$
(3) $C = \|T\|$
(5) The resulting operator T is unbounded, then …

(3.69)
(1) The most general Green function is $G(t) = \theta(t)\exp(-at) + c\exp(-at)$; $\exp(-at) \notin \mathscr{S}'$ but is a Schwartz distribution $\mathscr{D}'$
(2) $G_-(t) = A_-\exp(-at)$ for $t < 0$; $G_+(t) = A_+\exp(-at)$ for $t > 0$; imposing the discontinuity at $t = 0$ gives $A_+ = A_- + 1$, then $G(t) = A_-\exp(-at) + \theta(t)\exp(-at)$
(3–4) For the eq. in (1), $G(t) = \theta(t)\exp(-at) \in L^2(\mathbf{R})$ and is causal. The Green function belonging to $L^2(\mathbf{R})$ for the equation in (3) is "anticausal", i.e. $G(t) = 0$ if $t > 0$; the causal Green function is $G(t) = \theta(t)\exp(+at) \notin \mathscr{S}'$, as in (1)

(3.70)
(1) The limits are resp. $\theta(t)$ and $-\theta(-t) \in \mathscr{S}'$ and solve the equation in (1)(c).
(2) The most general solution is $x(t) = \theta(t) + c$; choosing suitably the arbitrary constant c one obtains the solutions in (1)(a)

(3.71)
(1)(a) The most general solution is $x(t) = \theta(t)\frac{\exp(-t)-\exp(-\alpha t)}{\alpha-1} + c\exp(-t)$, etc.
(b) $x(t) = t\exp(-t)\theta(t) + c\exp(-t)$
(2) $x(t) = \mathrm{sgn}\,t - 2\exp(-t)\theta(t) + c\exp(-t)$: the inverse Fourier transform can be obtained either by integration in the complex plane with indentation and careful use of Jordan lemma, or – more simply – using $2i\,\mathrm{P}\frac{1}{\omega}\frac{1}{1-i\omega} = 2i\,\mathrm{P}\frac{1}{\omega} - \frac{2}{1-i\omega}$

(3.72)
(1) $\widehat{G}(\omega) = i\,\mathrm{P}(1/\omega - 1) + c\,\delta(\omega)$ and $G(t) = \frac{1}{2}\exp(-i\,t)\mathrm{sgn}\,t + c\exp(-i\,t)$ (here, ∞^1 Green function are obtained, see q.(4))
(2) The most general solution is $x(t) = \frac{1+i}{4}\big(\mathrm{sgn}\,t\exp(-i\,t) - 2\exp(-t)\theta(t)\big) + c\exp(-i\,t)$ (a suggestion as in q.(2) of the above problem holds also in this case and in next question)
(3) $x(t) = i\big(\exp(-i\,t) - 1\big)\mathrm{sgn}\,t + c\exp(-i\,t)$

(3.73)
(1) $x(t) = \frac{a\sin t - \cos t}{a^2+1}$ (recall that $\delta(\omega-1)h(\omega) = \delta(\omega-1)h(1)$, etc.)
(2) (a) Using Parseval identity: $\|\widehat{x}_a(\omega)\|^2 \le \sup_\omega \frac{4\sin^2\omega}{a^2+\omega^2}\|\widehat{h}(\omega)\|^2 = C^2\|\widehat{h}(\omega)\|^2$; then, $C = 2/a$ as a rough estimate
(b) $C = 2$
(3) If $a \ne 0$, then $\widetilde{x}_a(t) \in L^2(\mathbf{R})$ and $C = 2/a$; but if $a = 0$, $\widetilde{x}(t) \notin L^2(\mathbf{R})$, in general

(3.74)
(1) $\|\widehat{x}(\omega)\|^2 \le \sup_\omega \frac{\omega^2}{b^2+a^2\omega^2}\|\widehat{f}(\omega)\|^2$, then $C = 1/a$
(3) $x(t) = \delta(t) - \theta(t)\ \exp(-t)\quad ;\quad x(t) = \theta(t)\ \exp(-t)$

(3.75)
(2) (a) $v^{(0)}(t) = \theta(t)\big(1-\exp(-t)\big) \notin L^2(\mathbf{R})$, but $\in \mathscr{S}'$
(b) $v^{(0)}(+\infty) = 1$ and the kinetic energy is $1/2$
(3) (a) $W_f = \frac{1}{2(1+\beta)} \to \frac{1}{2}$
$$(ii)\ (f,v) = \frac{1}{2\pi}(\widehat{f},\widehat{v}) = \frac{1}{2\pi}\int_{-\infty}^{+\infty}\frac{1}{1+\omega^2}\frac{1}{\beta-i\omega}\,d\omega = i\,\mathrm{R}(i) = \frac{1}{2(1+\beta)}$$

(3.76)
(2) $(f,v) = \frac{1}{2\pi}\int_{-\infty}^{+\infty}\frac{\widehat{f}^*(\omega)\widehat{f}(\omega)}{\beta-i\omega}\,d\omega \to \frac{\widehat{f}^2(0)}{2} = W_f^{(0)}$ (recall that $\widehat{f}^*(\omega) = \widehat{f}(-\omega)$)
(3) $v^{(0)}(+\infty) = \int_{-\infty}^{+\infty} f(t)\,dt = \widehat{f}(0)$
(4) $f(t)$ must be zero mean-valued

(3.77)
(1) $a = 2n\pi,\ n\in\mathbf{Z}$
(2) With $a=\pi$, one has $x(t) = (1/2)\big(\exp(-i\pi t)\big(\mathtt{sgn}(t-1)+\mathtt{sgn}(t)\big)\big) + c\exp(-iat)$
(3) $a = (2n+1)\pi,\ n\in\mathbf{Z}$, with $a=\pi$ one has
$x(t) = -(1/2)\big(\exp(-i\pi t)\big(+\mathtt{sgn}(t-1) - \mathtt{sgn}(t)\big)\big) + c\exp(-iat))$

(3.78)
(1) Via Fourier trasform one obtains in this case the most general Green function (precisely, ∞^2 Green functions). The causal Green function is $G(t) = \theta(t)\sin t \in \mathscr{S}'$
(2) $G_-(t) = A_-\cos t + B_-\sin t$ for $t<0$ and $G_+(t) = A_+\cos t + B_+\sin t$ for $t>0$. Imposing continuity at $t=0$ gives $A_- = A_+$, imposing the correct discontinuity of $\dot{G}(t)$ at $t=0$ gives $B_+ = B_- + 1$, then the most general Green function is $G(t) = A\cos t + B\sin t + \theta(t)\ \sin t$, as confirmed via Fourier transform.

(3.79)
Using Fourier transform one obtains in this case the most general Green function (precisely, ∞^2 Green functions), i.e., $G(t) = c + c_1 t + (1/2)|t|$. The Green function causal is $G(t) = t\,\theta(t) \in \mathscr{S}'$. As in the previous problem, one has $G_\mp(t) = A_\mp + B_\mp t$, and then $G(t) = A + B\,t + \theta(t)\,t$, which is the same as that obtained two lines above.

(3.80)
Differently from the two problems above, using Fourier transform one obtains *one* Green function for the first equation and ∞^1 Green functions for the second one.

The most general Green functions are obtained considering also the solutions of the homogeneous equation. For the first equation these are $A\exp(t) + B\exp(-t)$, which belong to $\mathscr{S}'$ only if $A = B = 0$. For the second equation, these solutions are $A + B\exp(\mp t)$, which belong to $\mathscr{S}'$ only if $B = 0$ (indeed, $\exp(\pm t) \in \mathscr{D}'$, the Schwartz distributions)

(3.81)
(1) The most general solution can be obtained via Fourier transform; one has $\widehat{x}(\omega) = D\mathrm{P}(1/\omega) - i\,\mathrm{P}(1/\omega) + 1/(1 - i\omega) + c_0\delta(\omega) + c_1\delta'(\omega)$ which gives

$$x(t) = \theta(t)\,\exp(-t) + (1/2)(t-1)\mathrm{sgn}\,t + c_0 + c_1 t$$

The solution respecting causality is

$$x(t) = \theta(t)\big(-1 + t + \exp(-t)\big)$$

(3.82)
Using Fourier transform one obtains only one solution; to have the most general solution one must add $c_1\exp(+t) + c_2\exp(-t)$. As in other similar cases (see, e.g., Problem 3.71, q.(2)), the general solution can be obtained either using Jordan lemma or by decomposition of $\widehat{x}(\omega)$ in simple fractions; in this case, the use of Jordan lemma is perhaps easier (simple poles at $\omega = \pm i$ and $\omega = -2i$). The solution respecting causality is

$$x(t) = \theta(t)\big(-(1/2)\exp(-t) + (1/3)\exp(-2t) + (1/6)\exp(+t)\big)$$

$\big(\notin L^2(\mathbf{R})$ and $\notin \mathscr{S}'\big)$; there is a solution in $L^2(\mathbf{R})$ given by

$$x(t) = -(1/6)\theta(-t)\exp(t) + \theta(t)\big(-(1/2)\exp(-t) + (1/3)\exp(-2t)\big)$$

(3.83)
Using Fourier transform one obtains only ∞^1 solutions; to have the most general solution one must add $c\,\exp(-t)$. One has $\widehat{x}(\omega) = \frac{-i}{(\omega-i)^2}\mathrm{P}\frac{1}{\omega} + c\,\delta(\omega)$.

To find $x(t)$, see the remark in the problem above; if the integration in the complex plane is adopted, see the suggestion in Problem 2.31.

The solution respecting causality is

$$x(t) = \theta(t)\big(1 - \exp(-t) - t\exp(-t)\big) \notin L^2(\mathbf{R})\,, \in \mathscr{S}'$$

(3.84)
The most general solution can be obtained via Fourier transform. One has $\widehat{x}(\omega) = \left(\mathrm{P}\frac{1}{\omega-1} - \mathrm{P}\frac{1}{\omega+1}\right)\frac{1/2}{1-i\omega} + c_1\delta(\omega-1) + c_2\delta(\omega+1)$ which gives (see the remark in the problems above; if Jordan lemma is used, two indentations are needed)

$$x(t) = (1/2)\theta(t)\,\exp(-t) + (1/4)(\sin t - \cos t)\mathrm{sgn}\,t + c'\cos t + c''\sin t$$

The solution respecting causality is

$$x(t) = (1/2)\theta(t)\big(\exp(-t) - \cos t + \sin t\big) \notin L^2(\mathbf{R}), \ \in \mathscr{S}'$$

(3.85)
(1) (a) $x(t) = c\exp(-it) + i\exp(-i\alpha t)/(\alpha - 1)$
(b) $x(t) = (t + c)\exp(-it)$
(2) $A = -2$. In the case $\alpha = 1$, $x(t) = c_1\cos t + c_2\sin t - (t/2)\cos t$

(3.86)
(1)(b) $x(t) = f_n(t) + c_1\cos t + c_2\sin t$
(2) $g_n(t) = f_n(t)$
(3) 0 and $i\pi\big(\delta(\omega - 1) - \delta(\omega + 1)\big)$

(3.87)
(1) $k^2\widehat{u}(k) = \exp(ikx_1)$, recall that $\mathscr{F}^{-1}\big(D\mathrm{P}(1/k)\big) = (1/2)|x|$, then $u(x, x_1) = -(1/2)|x - x_1| + c_0 + c_1 x$; the boundary conditions give $c_0 = x_1/2$, $c_1 = (1/2) - x_1$, etc.
(2) $u_\mp = A_\mp x + B_\mp$ resp. for $0 \le x < x_1$ and $x_1 < x \le 1$; the four constants $A_\mp$, $B_\mp$ are determined imposing $u_-(0) = u_+(1) = 0$, $u_-(x_1) = u_+(x_1)$ and $u'_+(x_1) = u'_-(x_1) - 1$
(3) $u'(x) = -\theta(x - x_1) + c_1$, etc.

The solution is $u(x, x_1) = \begin{cases} x(1 - x_1) & \text{for} \quad 0 \le x \le x_1 \\ x_1(1 - x) & \text{for} \quad x_1 \le x \le 1 \end{cases}$

(3.88)
The most general solution is $u(x) = (1/2)\big(|x| - \mathrm{sgn}\,x\big) + \theta(x)\exp(-x) + c_1 + c_2 x$; the solution vanishing at $x \to +\infty$ is $u(x) = \theta(-x)(1 - x) + \theta(x)\exp(-x)$

(3.89)
Apart from the notations, the general solution is given in Problem 3.84. Notice that the solution is continuous at $x = 0$. The solution satisfying the given boundary conditions is $u(x) = (1/2)\big(\exp(-x) - \cos x - \exp(-\pi/2)\sin x\big)$. To explain the non-existence of solutions with the conditions $u(0) = u(\pi) = 0$, observe that the problem is equivalent to that of solving the equation $Tu = f(x)$ where $T = d^2/dx^2 + I$ is the operator in $L^2(0, \pi)$ with given boundary conditions; as seen in Sect. 1.2.5, equations of this form may admit no solution: see for instance Problems 1.92, 1.93, 1.95

(3.90)
(1) Writing $\mathscr{F}\big(G(t)\big) = \widehat{G}(\omega) = A(\omega)\exp\big(i\Phi(\omega)\big)$, one has

$$\widehat{b}(\omega) = 2\pi\,\widehat{G}(\omega)\,\delta(\omega - \omega_0) = 2\pi\,A(\omega_0)\exp(i\Phi(\omega_0))\,\delta(\omega - \omega_0)$$

then $b(t) = A(\omega_0)\exp\big(-i\omega_0 t + i\Phi(\omega_0)\big)$

(2) The output is the superposition of two waves with different amplitudes and a "phase distortion" (in general):

$$b(t) = \exp(i\Phi_1)\Big(A_1 \exp(-i\omega_1 t) + A_2 \exp\big(-i\omega_2 t + i(\Phi_2 - \Phi_1)\big)\Big)$$

where $\Phi_1 = \Phi(\omega_1)$, $A_1 = A(\omega_1)$ etc.; in general, $A_1 \neq A_2$, $\Phi_1 \neq \Phi_2$

(3.91)
(1) The Fourier transform $\widehat{f}(\omega)$ of a real function $f(t)$ has the property $\widehat{f^*}(\omega) = \widehat{f}(-\omega)$, then the condition is: n even
(2) $b(t) = -(1/2)\cos t$
(3) $\widehat{b}(\omega) = \left(\mathrm{P}\frac{1}{\omega+1} - \mathrm{P}\frac{1}{\omega-1}\right)\frac{1}{(\omega-i)^2}$;
to find $b(t)$, see the remarks in Problems 3.83 and 2.31. One obtains
$b(t) = (t-1)\exp(t)\,\theta(-t) - (1/2)\cos t\ \mathrm{sgn}\,t$

(3.92)
(1)(a) $\widehat{b}(\omega) = -i\omega\widehat{a}(\omega)$, then $b(t) = da(t)/dt$ and $b(t) = \delta(t)$
(2) $b_\varepsilon(t) = (t/\varepsilon^2)\exp(-t/\varepsilon)\theta(t) \to \delta(t)$

(3.93)
(1–2) $\widehat{a}(\omega) = i\,\mathrm{P}(1/\omega) + 3\pi\delta(\omega)$, $\widehat{b_\tau}(\omega) = \frac{i\omega\exp(i\omega\tau)}{(1+\omega^2)^2} \in L^2(\mathbf{R})$ and is infinitely differentiable in $L^1(\mathbf{R})$, then $b_\tau(t) \in L^2(\mathbf{R})$, is rapidly vanishing as $|t| \to \infty$ and continuously differentiable because $\omega\,\widehat{b_\tau}(\omega) \in L^1(\mathbf{R})$
(3) Considering $b_\tau(t)$ as distributions, $b_\tau(t)$ converges to 0 in $\mathscr{S}'$, the same for $\widehat{b_\tau}(\omega)$, indeed $<\widehat{b_\tau}, \varphi> = <\exp(i\omega\tau), \ldots>$, $\forall\varphi \in \mathscr{S}$, can be seen as the Fourier transform of a function $\in L^1(\mathbf{R})$ and then $\to 0$ thanks to Riemann-Lebesgue theorem. Exactly the same argument shows the weak convergence in $L^2(\mathbf{R})$, it is enough to consider the scalar product in $L^2(\mathbf{R})$ $\big(\widehat{b_\tau}(\omega), f(\omega)\big) = \big(\exp(i\omega\tau)\frac{i\omega}{(1+\omega^2)^2}, f(\omega)\big)$ $\forall f \in L^2(\mathbf{R})$. Clearly, no convergence in the $L^2(\mathbf{R})$ norm

(3.94)
(1–2) $b_T(t) = (1/2)\big(\exp\big(-(t-T)^2\big) - \exp(-t^2)\big) \to -(1/2)\exp(-t^2) = b_\infty(t)$ and $\widehat{b_T}(\omega) = \big(1 - \exp(i\omega T)\big)\widehat{b_\infty}(\omega) \to \widehat{b_\infty}(\omega)$.
The limit is in $\mathscr{S}'$, and in the weak $L^2(\mathbf{R})$ convergence: see problem above, q. (3)
(3) Expectedly, the limit coincides with the result in (2)

(3.95)
(1) Yes: $b_c(t) = 0$, $\forall t$, if the support of the $\mathscr{F}$ transform $\widehat{a}(\omega)$ has no intersection with the interval $|\omega| < c$
(2) In general $\widehat{b_c}(\omega)$ is not a continuous function, then in general $b_c(t) \notin L^1(\mathbf{R})$, but $\in L^2(\mathbf{R})$, is bounded, (possibly not rapidly) vanishing at $|t| \to \infty$, infinitely differentiable
(3) $b_c(t) \to a(t)$, in norm $L^2(\mathbf{R})$, indeed $\|\widehat{a}(\omega) - \widehat{b_c}(\omega)\|^2 = \int_{|\omega|>c} |\widehat{a}(\omega)|^2\, d\omega \to 0$

(3.96)
(1) The only admitted $a(t)$ must satisfy $\widehat{a}(\omega) = \sum_{0\neq n\in\mathbf{Z}} c_n \delta(\omega - \frac{\pi n}{T}) \in \mathscr{S}'$ (not $\in L^2(\mathbf{R})$), see Problem 3.57 *(a)*
(2) In general $\widehat{b_T}(\omega)$ is not a continuous function, then in general $b_T(t) \notin L^1(\mathbf{R})$, but $\in L^2(\mathbf{R})$, is bounded, (possibly not rapidly) vanishing at $|t| \to \infty$, continuous but in general not differentiable
(4) $\widehat{b_T}(\omega) \to 2\pi\delta(\omega)\widehat{a}(\omega) = 2\pi\widehat{a}(0)\delta(\omega)$, $b_T(t) \to \widehat{a}(0) = \int_{-\infty}^{+\infty} a(t)\,dt$

(3.97)
(1) $a(t) = b(t) + \alpha\,\theta(t)\exp(-t) * b(t)$ then $\widehat{a}(\omega) = \frac{1-i\omega+\alpha}{1-i\omega}\widehat{b}(\omega)$
If $\alpha = 1:\ G(t) = \delta(t) - \theta(t)\exp(-2t)$; if $\alpha = -1:\ G(t) = \delta(t) + \theta(t) +$ const;
if $\alpha = -2:\ G(t) = \delta(t) - 2\theta(-t)\exp(+t)$
(2) $\beta = 1$
(3) The causal Green function is $G(t) = \delta(t) + 2\theta(t)\exp(+t) \notin \mathscr{S}'$

(3.98)
(1) $G(t) = \begin{cases} 1 & \text{for } 0 < t < 1 \\ 0 & \text{elsewhere} \end{cases} + \text{const}$
(2) With $a(t) = \theta(t)$, one obtains $b(t) = t\,\theta(t) - (t-1)\,\theta(t-1) + c$, see also next question
(3) The most general Green function is $G(t) = -\theta(-t) + \theta(t+1) + c$
Using Fourier transform, with $a(t) = \theta(t)$, one has
$-i\,\omega\,\widehat{b}(\omega) = i\,\mathrm{P}(1/\omega)\big(1 + \exp(i\omega)\big) + 2\pi\,\delta(\omega)$,
see now Problem 3.58, *(a)* and *(d)*; but perhaps the direct integration is simpler …:
$b(t) = t\,\theta(t) + (t-1)\,\theta(t-1) + c$

(3.99)
(1) The Green function belonging to $L^2(\mathbf{R})$ is $G_0(t) = \begin{cases} -t-1 & \text{for } -1 < t < 0 \\ t-1 & \text{for } \ 0 < t < 1 \\ 0 & \text{elsewhere} \end{cases}$
The most general is $G_0(t) + c_0 + c_1 t$
(2) With $a(t) = \dot{\delta}(t)$, see Problem 3.29, q.(1);
With $a(t) = t$, a solution is $\widehat{b}(\omega) = \widehat{G}_0(\omega)\big(-2\pi i\,\delta'(\omega)\big)$ and then
$b(t) = -i < \delta'(\omega), \widehat{G}_0(\omega)\exp(-i\omega t) >= i < \delta(\omega), \frac{d}{d\omega}\ldots >= t$
and the most general $b(t) = c_0 + c_1 t$

(3.100)
(a) $b(t) = a(-t)$
(b) Using $\mathscr{F}\big(a(-t)\big) = \widehat{a}(-\omega)$, one obtains $\widehat{b}(\omega) = \frac{1}{(1-i\omega)(\alpha+i\omega)}$,
then $b(t) = \frac{1}{\alpha+1}\big(\exp(\alpha t)\theta(-t) + \exp(-t)\theta(t)\big)$
(2) $G(t,t') = M(t')\delta(t-t')$

(3.101)
(1) $\mathscr{F}^{-1}(-\mathrm{P}(1/\omega)*g)=\pi i(\mathrm{sgn}\,t)f(t)=\pi i f(t)$, then $f(t)=0$ if $t<0$
(2–3) Only one satisfies the property given in (1), in agreement with Problem 3.7, q.(1).

Notice that $\exp(+i\omega)$ in the upper half complex plane $\omega=\omega_1+i\omega_2$ becomes $\exp(+i\omega_1)\exp(-\omega_2)$ which vanishes for $\omega_2\to+\infty$ $\big($whereas $\exp(-i\omega)$ diverges$\big)$
(4) $\widehat{G}_2(\omega)=\dfrac{\omega}{1+\omega^2}$ and $G(t)=\theta(t)\,\exp(-t)$.

(3.102)
(1–2) The ODE for $\widehat{u}(k,t)$ is $\frac{d\widehat{u}}{dt}=-k^2\widehat{u}$,
then $\widehat{u}(k,t)=\widehat{f}(k)\exp(-k^2t)=\widehat{f}(k)\widehat{G}(k,t)$ where $\widehat{f}(k)=\mathscr{F}\big(f(x)\big)$. The Green function is $G(x,t)=\exp(-x^2/4t)/2\sqrt{t\pi}$
(3) The limit is $\delta(x)$, which can be obtained either from $\lim_{t\to0}\exp(-k^2t)=1$ or directly from $G(x,t)$ $\big($see Problem 3.23, q.(3): replace n^2 with $1/4t\big)$.

This agrees with the result $f(x)*\delta(x)=f(x)$
(4) From $\exp(-k^2)\widehat{f}(k)=\exp(-1)\widehat{f}(k)$ one has $\widehat{f}(k)=c'\delta(k-1)+c''\delta(k+1)$, then $f(x)=c_1\cos x+c_2\sin x\in\mathscr{S}'$

(3.103)
(1) Use Fourier transform: $\widehat{T_t}\,\widehat{u}(0)=\exp(-k^2t)\widehat{u}(0)$, and $\|T_t\|=\|\widehat{T_t}\|=1$
(2) $\|\widehat{T_t}\widehat{f}(k)\|^2=\int_{-\infty}^{+\infty}\exp(-2k^2t)|\widehat{f}(k)|^2\,dk\to0$, use Lebesgue theorem, indeed $|\widehat{f}(k)|^2\in L^1(\mathbf{R})$, etc., then: strong convergence (not norm convergence) to zero
(3) $\|(\widehat{T_t}-I)\widehat{f}(k)\|^2=\int_{-\infty}^{+\infty}|\exp(-k^2t)-1|^2|\widehat{f}(k)|^2\,dk\to0$

but $\|\widehat{T_t}-I\|=\sup_{k\in\mathbf{R}}|\exp(-k^2t)-1|=1$, for any $t\geq0$, then strong convergence (not norm convergence) to the identity operator

(3.104)
(1) $u(x,t)\in L^2(\mathbf{R})$, vanishes as $x\to\pm\infty$ but not rapidly, and $\notin L^1(\mathbf{R})$, indeed $\widehat{u}(k,t)$ is not a continuous function
(2) All the functions as $\exp(-k^2t)\widehat{f}(k)$, $k^n\exp(-k^2t)\widehat{f}(k)$, $k^n\exp(-k^2t)$, $\forall n=0,1,2,\ldots$, etc., belong to $L^1(\mathbf{R})$, then ...
(3) $u(x,t)=1$; $u(x,t)=x$; $u(x,t)=x^2+2t$.

For instance, in the case $f(x)=x$, one has $\mathscr{F}(x)=-2\pi i\delta'(k)$, and then $u(x,t)=(1/2\pi)(-2\pi i)\int_{-\infty}^{+\infty}\delta'(k)\exp(-k^2t-ikx)\,dk=i<\delta(k),\frac{d}{dk}\ldots>=x$

(3.105)
(1) $\widehat{u}(k,t)=\widehat{f}(k)\exp(-k^2t+iakt)$ and $G(x,t)=\dfrac{\exp\big(-(x-a)^2/4t\big)}{2\sqrt{\pi t}}$
(2) $\|\widehat{u}_a(k,1)-\widehat{u}_0(k,1)\|^2=\int_{-\infty}^{+\infty}\exp(-2k^2)|\widehat{f}(k)|^2|\exp(iak)-1|^2\,dk\to0$
(3) Yes: if $f(x)=\delta''(x)$, one has $|\widehat{f}(k)|^2=k^4$ and the above result still holds
(4) $\widehat{u}_0(k,1)=2i\exp(-k^2)\mathrm{P}(1/k)\notin L^2(\mathbf{R})$, but

$\widehat{u}_a(k,1)-u_0(k,1)=2i\exp(-k^2)\Big(\frac{\exp(ika)-1}{k}\Big)\in L^2(\mathbf{R})$ (the symbol P can be omitted), etc.

(3.106)
(1) The ODE for $\widehat{u}(k,t)$ is $\widehat{u}_{tt}=-k^2\widehat{u}$, then $\widehat{u}(k,t)=A(k)\cos kt+B(k)\sin kt$
(2) If $\widehat{u}_t(k,0)=0$ one obtains $\widehat{u}(k,t)=\widehat{f}(k)\cos kt=(1/2)\widehat{f}(k)\big(\exp(ikt)+\exp(-ikt)\big)$ and then $u(x,t)=(1/2)\big(f(x-t)+f(x+t)\big)$
(3) $\widehat{f}(k)\exp(\pm kit)\to 0$ in $\mathscr{S}'$ because $<\exp(\pm kit),\widehat{f}(k)\varphi(k)>\to 0$ thanks to the Riemann-Lebesgue theorem $\big(\widehat{f}(k)\varphi(k)\in L^1(\mathbf{R})\big)$. The same argument shows that the limit holds also in the weak $L^2(\mathbf{R})$ convergence
(4) The limit is $f(x)=u(x,0)$ in the sense of the $L^2(\mathbf{R})$ norm:
$\|\widehat{f}(k)(\cos kt-1)\|^2=\int_{-\infty}^{+\infty}\ldots\to 0$, thanks to Lebesgue theorem

(3.107)
(1) $\widehat{G}(k,t)=\frac{\sin kt}{k}$ and $G(x,t)=\begin{cases}1/2 & \text{for } |x|<t\\ 0 & \text{for } |x|>t\end{cases}$
(2) $\widehat{G}_t(k,t)=\cos kt,\quad G_t(x,t)=(1/2)\big(\delta(x+t)+\delta(x-t)\big)$. $\big($*Warning*: it is requested to find the derivative with respect to t, not to x: this may be obtained either directly from $G(x,t)$, or from $\widehat{G}(k,t)\big)$.
$G(x,t)$ is indeed the solution if $g(x)=\delta(x)$
(3) $\widehat{u}(k,t)$ has a simple pole in $k=+i$, then, for $t-x<0$ one has $u(x,t)=0$, according to Jordan lemma. Indeed, there is a wave propagating along x with velocity 1.

(3.108)
(1) The ODE $-k^2\widehat{u}(k,t)-c^{-2}\widehat{u}_{tt}(k,t)=\exp(ikvt)$ admits the particular solution $k^2\widehat{u}(k,t)=\frac{c^2}{v^2-c^2}\exp(ikvt)$ which gives
$$u(x,t)=\frac{c^2}{2(v^2-c^2)}|x-vt| \text{ plus arbitrary functions } f(x-ct)+g(x+ct)$$
(2) The ODE admits the particular solution $k\,\widehat{u}(k,t)=(ivt/2)\exp(ikvt)$ which gives
$u(x,t)=(vt/4)\,\mathrm{sgn}(x-t)$ + etc., as in (1)

(3.109)
(1) $\widehat{u}(k,t)=\big(\widehat{f}(k)(1-ikt)+t\,\widehat{g}(k)\big)\exp(ikt)$
(2) No, because $\widehat{f}(k)\in L^2(\mathbf{R})$ does not imply $k\,f(k)\in L^2(\mathbf{R})$, see q.(4)
(3) $u(x,t)=f(x-t)+t\,f_x(x-t)+t\,g(x-t)$
(4) $u(x,t)=(1-t)\theta(x-t)\exp(t-x)+t\,\delta(x-t)$

(3.110)
(1) The ODE for $\widehat{u}(k,y)$ is $\frac{d^2\widehat{u}}{dy^2}-k^2\widehat{u}=0$
(2) $G(x,y)=(y/\pi)/(x^2+y^2)$
(3) The limit is $\delta(x)$, indeed, $u(x,0)=f(x)*\delta(x)=f(x)$

(3.111)
(1) $u(x,y)$ is the real part of an analytic function
(2) $\widehat{u}(k,y)=\pi\exp\big(-|k|(y+1)\big)$, then ..., see Problem 2.45, q.(3)
(3) If $f(x)=\sin x$ one obtains $u(x,y)=\sin x\,\exp(-y)$

The case $f(x) = x$ is critical: using $\widehat{u}(k, y) = A'(k)\exp(-|k|y)$ according to the problem above, one obtains the unexpected difficulty
$u(x, y) = (1/2\pi)(-2\pi i)\int_{-\infty}^{+\infty}\delta'(k)\exp(-|k|y - ikx)\,dk = i < \delta(k), \frac{d}{dk}\ldots >$
$= x - \ldots$. Actually, in this case where the support of $\widehat{f}(k)$ is in the single point $k = 0$, there is no reason to exclude the term $B'(k)\exp(|k|y)$. See next problem: the general solution is indeed $u(x, y) = x + c\,y$ with arbitrary c

(3.112)
(2) The solutions are proportional to y, xy, etc.

(3.113)
(1) $\widehat{f}(k_1, k_2, k_3) = -8i\pi^2\big(1/(1 + k_1^2)\big)\delta'(k_2)\delta(k_3)$
(2)

$$\mathscr{F}\big(f(r)\big) = \widehat{f}(k) = 2\pi\int_0^{+\infty}\int_{-1}^{1}\exp(ikru)\,r^2 f(r)\,du\,dr = 4\pi\int_0^{+\infty}\frac{\sin kr}{kr}r^2 f(r)\,dr$$

(where $u = \cos\theta$). With $f_1(r) = 1/r^2$, the result is (see Problem 2.28)
$\mathscr{F}(1/r^2) = \dfrac{2\pi}{k}\displaystyle\int_{-\infty}^{+\infty}\frac{\sin\xi}{\xi}\,d\xi = \frac{2\pi^2}{k}$
With $f_2(r) = \exp(-r)$, the result is $\mathscr{F}\big(\exp(-r)\big) = \dfrac{8\pi}{(1 + k^2)^2}$

(3.114)
(1) $\mathscr{F}(1/r) = 4\pi/k^2$; recall that $\mathscr{F}^{-1}(g(\mathbf{k})) = (1/2\pi)^3\int_{\mathbf{R}^3}\exp(-i\,\mathbf{k}\cdot\mathbf{x})g(\mathbf{k})\,d^3\mathbf{k}$
(2) $-k^2\widehat{V}(k) = -4\pi\,\widehat{\rho}(k)$ then $\widehat{G}(k) = 4\pi/k^2$ and $G(r) = 1/r$, which is the potential due to a point charge, as well-known

(3.115)
In the first case: $-4\pi^2(-k_1^2 - k_2^2)\big(\delta''(k_1)\delta(k_2) - \delta(k_1)\delta''(k_2)\big) = 0$

(3.116)
(1) $x^\alpha < M\exp(\varepsilon x)$, for any $\varepsilon > 0$ "arbitrarily small", then $\lambda = 0$;
$x^\alpha\exp(-x^2) < M\exp(-ax)$ for any $a > 0$ "arbitrarily large", then $\lambda = -\infty$;
for the last function: $\lambda = \beta$; if $\alpha \le -1$ the functions are not (locally) summable
(2) $\frac{\exp(\gamma x)}{x+c} < M\exp(ax)$ for any $a > \mathrm{Re}\,\gamma$, then $\lambda = \mathrm{Re}\,\gamma$;
the other transforms have $\lambda = 0$. If $c \le 0$ the functions are not (locally) summable

(3.117)
(1) Simple pole at $s = -1$, essential singularity at $s = \infty$
(2) $f(x) = \begin{cases} 2 - x - 2\exp(-x) & \text{for } 0 < x < 1 \\ (e - 2)\exp(-x) & \text{for } x > 1 \end{cases}$
(3) $\lambda = -1$, in agreement with (1) and (2)

(3.118)
$f(x) = \sin x$ for $0 \le x \le \pi$ and $= 0$ elsewhere; $\lambda = -\infty$

(3.119)
(1) $\widehat{f}(\omega)$ is the function of $\omega \in \mathbf{R}$ obtained replacing s with $-i\omega$ in $\widetilde{f}(s)$
(2) $f(x) = x$ if $0 \le x < 1$ and $= 0$ elsewhere, with $\lambda = -\infty$ (see Problem 3.46, q.(2))
(3) $\mathscr{F}\big(\theta(x)\big) = \lim_{\varepsilon \to 0^+} \frac{1}{\varepsilon - i\omega} = i\,\mathrm{P}\frac{1}{\omega} + \pi\delta(\omega)$; $\mathscr{L}\big(\theta(x)\big) = 1/s$
(4)

$$f(x) = \begin{cases} x & \text{for } 0 \le x \le 1 \\ 1 & \text{for } x \ge 1 \end{cases}$$

(and $f(x) = 0$ if $x \le 0$, of course); with $\lambda = 0$. The Fourier transform is (see Problem 3.46, q.(4))

$$\widehat{f}(\omega) = \Big(\frac{\exp(i\omega) - 1}{\omega}\Big)\mathrm{P}\Big(\frac{1}{\omega}\Big) + \pi\delta(\omega)$$

(3.120)
All these functions satisfy the summability criterion $|f(x)| \le M \exp(\varepsilon x)$ for any $\varepsilon > 0$ "arbitrarily small", then $\lambda = 0$. The first three admit also Fourier transform (resp. in L^1, L^2, $\mathscr{S}'$); $\theta(x)\exp(\sqrt{x}) \notin \mathscr{S}'$

(3.121)
(1) Let $f(x)$ have compact support. Up a translation (when necessary), which introduces a trivial phase factor $\exp(i\omega a)$ in the Fourier transform, the support can be placed in $x \ge 0$. The Laplace transform of this function is then analytic in all $s \in \mathbf{C}$ and the Fourier transform analytic in all $\omega \in \mathbf{C}$
(3) The Laplace transforms of these functions, thanks to the summability criterion, have $\lambda = -\infty$ then …

(3.122)
(1) $a - b = 2n\pi,\ n \in \mathbf{Z}$
(3) With $n > m$:
$f_{nm}(x) = \mathscr{F}^{-1}\big(g_{nm}(\omega)\big) = \sin(x-1)\big(\theta(x - 1 - 2m\pi) - \theta(x - 1 - 2n\pi)\big)$
The support is $1 + 2\pi m \le x \le 1 + 2\pi n$

(3.123)

$$\widetilde{G}(s) = \frac{s\,C}{s^2 LC + s\,RC + 1}$$

(3.124)
(2) Putting $Y(s) = \mathscr{L}\big(y(t)\big)$, one has $Y(s) = \dfrac{1 - \exp(-sc)}{s(s^2 + 1)}$

With $c = \pi$, the solution is $y(t) = \begin{cases} 1 - \cos t & \text{for } 0 \le t \le \pi \\ -2\cos t & \text{for } t \ge \pi \end{cases}$

With $c = 2\pi$ the solution is $y(t) = 1 - \cos t$ for $0 \le t \le 2\pi$ and $= 0$ elsewhere

If $c = 2n\pi$, $n = 1, 2, \ldots$, then $y(t) = 0$ for any $t \ge c$

(3) With $c = 2\pi$ the solution is $y(t) = \begin{cases} \sin t & \text{for } 0 \le t \le \pi \\ 2\sin t & \text{for } t \ge \pi \end{cases}$

If $c = (2n-1)\pi$, $n = 1, 2, \ldots$, then $y(t) = 0$ for any $t \ge c$

(3.125)
$\widetilde{f}(s) = \log\big((s-b)/(s-a)\big)$; $\widetilde{f}(s) = (\pi/2) - \arctan s$

(3.126)
(1) From $\mathscr{L}(x\, y'') = -(d/ds)\big(s^2 \widetilde{J}_0(s) - s y_0 - y_0'\big) = \ldots$

one obtains $(s^2+1)\dfrac{d\,\widetilde{J}_0(s)}{ds} + s\,\widetilde{J}_0(s) = 0$, then $\widetilde{J}_0(s) = 1/\sqrt{1+s^2}$

(2) $J_0(x) * J_0(x) = \theta(x)\sin x$

(3.127)
(a) $\mathscr{L}\big(f(x)\big) = F_0(s)/\big(1 - \exp(-sT)\big)$

For the square wave: $\mathscr{L}\big(f(x)\big) = \dfrac{\big(1-\exp(-s)\big)^2}{s\big(1-\exp(-2s)\big)} = \dfrac{1}{s}\tanh(s/2)$

(b) $f(x) = \theta(x)|\sin x|$

(c) $f(x) = n$ for $n-1 < x < n$, $(n = 1, 2, \ldots)$, and $f(x) = 0$ for $x < 0$, of course

(3.128)
(1) The ODE is $\dfrac{d^2\widetilde{u}}{dx^2} - s^2\widetilde{u} = -s\, f(x) - g(x)$

(2) $\widetilde{u}(x,s) = A(s)\exp(sx) + B(s)\exp(-sx)$

(3) The boundedness of the solution $\forall x > 0$ implies $A(s) = 0$, then $\widetilde{u}(0,s) = B(s) = \mathscr{L}\big(\varphi(t)\big) = \widetilde{\varphi}(s)$ and $\widetilde{u}(x,s) = \widetilde{\varphi}(s)\exp(-sx)$, then finally $u(x,t) = \theta(t-x)\varphi(t-x)$, which describes the propagation of a wave along the string with velocity 1, produced by the displacement at $x = 0$

(3.129)
(1) Apply a "rotated" Jordan lemma in the complex plane s, with $s = s_1 + i\, s_2$ and $\exp(+sx) = \exp(+s_1 x)\exp(i s_2 x)$: if one looks for the inverse Laplace transform $f(x)$ for $x < 0$, the integral along the "vertical" line ℓ must be completed with a semicircle in the right half plane $s_1 > \lambda$, where $\widetilde{f}(s)$ is analytic, which gives $f(x) = 0$ for $x < 0$, as expected

(2) Looking for $f(x)$ for $x > 0$, the integral along the "vertical" line ℓ must be completed with a semicircle in the left half plane. In this case, two single poles $s = \pm i$ are present

(3.130)
Proceeding as in previous problem, q. (2), in the absence of other singularities apart from the cut, the integral along the vertical line is transformed into the integrals along the upper and lower margins of the cut (cf. Problem 2.35):

$$f(x) = \int_{\ell} \ldots = -\frac{1}{2\pi i} 2 \int_{-\infty}^{0} \frac{\exp(s_1 x)}{i\sqrt{|s_1|}}\, ds_1 = \frac{1}{2\pi} \int_{-\infty}^{+\infty} \exp(-u^2 x)\, du = 1/\sqrt{\pi x}$$

(3.131)
(1) With $s_1 = \mathrm{Re}\, s > 0$, the Laplace transform of a function $f(x) \in L^2(0, +\infty)$ can be seen as a scalar product in $L^2(\mathbf{R})$ (assume for simplicity $f(x)$ real) $\mathscr{L}\big(f(x)\big) = \big(f(x), \exp(-sx)\big) = \widetilde{f}(s)$, then $\lambda \le 0$ and $\widetilde{f}(s)$ is analytic (at least) in the half plane $s_1 > 0$. The completeness condition $\big(x^n \exp(-x), f(x)\big) = 0,\ \forall n = 0, 1, \ldots$ becomes the condition on the derivatives $\left.\dfrac{d^n \widetilde{f}(s)}{ds^n}\right|_{s=1} = 0$ of the Laplace transform $\widetilde{f}(s)$ evaluated at the point $s = 1$. But an analytic function vanishing in a point with all its derivatives is zero
(2) The completeness condition $\big(\exp(-x)\exp(-x/n), f(x)\big) = 0$ becomes $\widetilde{f}(1 + \frac{1}{n}) = 0$, this produces a sequence of zeroes accumulating at the point $s = 1$, which is an analyticity point for the Laplace transform $\widetilde{f}(s) = \mathscr{L}\big(f(x)\big)$: this implies $\widetilde{f}(s) \equiv 0$ (see Problem 2.2)

Problems of Chap. 4

(4.1)
The cosets are disjoint: indeed, assume that some element belongs to two different cosets $g_1 H$ and $g_2 H$, i.e., assume that $g_1 h_1 = g_2 h_2$ for some h_1, h_2; this implies $g_2 = g_1 h'$, then $g_2 \in g_1 H$, and $g_2 H = g_1 H$. Also, different elements $h_1, h_2 \in H$ give different elements in the coset gH: indeed, $gh_1 = gh_2$ clearly implies $h_1 = h_2$, etc.

(4.2)
Let $v_1 \in V_1,\ v_2 \in V_2 = V_1^{\perp},\ g \in G$; by hypothesis, $\mathscr{R} : V_1 \to V_1$, then

$$0 = \big(v_2, \mathscr{R}(g)v_1\big) = \big(\mathscr{R}^{+}(g)v_2, v_1\big) = \big(\mathscr{R}^{-1}(g)v_2, v_1\big) = \big(\mathscr{R}(g^{-1})v_2, v_1\big)$$

therefore $\mathscr{R}(g^{-1})v_2$ is orthogonal to v_1, for any $v_1 \in V_1,\ v_2 \in V_2 = V_1^{\perp},\ g \in G$, this gives just $\mathscr{R} : V_2 \to V_2$

(4.3)
(1)*(i-ii)* One has $Tv' = T\mathscr{R}(g)v_\lambda = \mathscr{R}(g)Tv_\lambda = \lambda\big(\mathscr{R}(g)v_\lambda\big) = \lambda\, v'$, then V_λ is an invariant subspace: $T : V_\lambda \to V_\lambda$. An irreducible representation $\mathscr{R}$ has no (proper) invariant subspace, then $V_\lambda \equiv V$ and $T = \lambda\, I$
(2) Start with a representation $\mathscr{R}$ sum of two inequivalent irreducible representations; choose a basis in V where $\mathscr{R}$ takes the form of a matrix with two diagonal blocks, then write (with clear notations)

$$\mathscr{R} = \begin{pmatrix} \mathscr{R}_1 & 0 \\ 0 & \mathscr{R}_2 \end{pmatrix} \quad \text{and} \quad T = \begin{pmatrix} T_1 & A_1 \\ A_2 & T_2 \end{pmatrix} \quad \text{etc.}$$

(4.4)
(1) Apply Schur lemma (previous problem) replacing the operator T with any element $\mathscr{R}(g)$ of the given representation; one gets that each $\mathscr{R}(g)$ leaves any 1- dimensional subspace invariant, ... An alternative proof: in the case of Abelian groups, each conjugacy class contains just an element, then $s = N$, where N is the order of the group, and recalling also Burnside theorem, ...
(2) It is known, and can be easily verified, that the kernel K of any group homomorphism (in particular, of any representation of the group) is an invariant subgroup of the group G. By definition, a simple group does not contain (proper) invariant subgroups; then, apart from the trivial case where the kernel is the whole group (which corresponds to the trivial representation $\mathscr{R} : g \to 1,\ \forall g \in G$), the representations of a simple group are faithful. Apart from the trivial one, all the irreducible representations of O_2 are 2-dimensional

(4.5)
(1) The group contains 6 elements, separated into 3 conjugacy classes: the identity, the rotations, the 3 reflections; there are 3 inequivalent irreducible representations, two of dimension 1, and one of dimension 2, in agreement with Burnside theorem: $6 = 1^2 + 1^2 + 2^2$, only one is faithful. The expected degeneracies are then 1 and 2
(2) The characters are (the first two are obvious, ...)

$$\chi_{\mathscr{R}_1} = (1;\, 1,\, 1;\, 1,\, 1,\, 1) \quad \chi_{\mathscr{R}'} = (1;\, 1,\, 1;\, -1 - 1,\, -1) \quad \chi_{\mathscr{R}_2} = (2;\, -1,\, -1;\, 0,\, 0,\, 0)$$

where the elements of the group are written in the order: identity, rotations, reflections, and where $\mathscr{R}_1$ denotes the trivial representation, $\mathscr{R}'$ the other 1-dimensional representation, and $\mathscr{R}_2$ the 2-dimensional one

(4.6)
(1) The 6-dimensional representation $\mathscr{R}$ decomposes in this way: $\mathscr{R} = \mathscr{R}_1 \oplus \mathscr{R}' \oplus 2\,\mathscr{R}_2$, with the notations as in the previous problem. $\mathscr{R}', \mathscr{R}_2$ describe resp. the rigid rotations and the rigid displacements of the system; the two other representations describe oscillations, one of these is then doubly degenerate

(2) The "surviving" symmetry is that of the isosceles triangle, which admits only 1-dimensional representations. Then, Schur lemma indicates that the doubly degenerate frequency should separate into two distinct frequencies

(4.7)
$N = 8$ with 5 conjugacy classes. There are four 1-dimensional representations and one 2-dimensional representation: $8 = 1^2 + 1^2 + 1^2 + 1^2 + 2^2$. The degeneracies are then 1 and 2. But, e.g.., with the notations of Problem 1.101, the eigenvalue $\lambda = -50$ has degeneracy 3, being obtained when $n = 1,\ m = 7;\ n = 7,\ m = 1$ and $n = m = 5$

(4.8)
There are two 1-dimensional representations, one of dimension 2, and two of dimension 3 of $\mathcal{O}_1$, in agreement with Burnside theorem: $24 = 1^2 + 1^2 + 2^2 + 3^2 + 3^2$ (resp. four of dimension 1, two of dimension 2, and four of dimension 3 of $\mathcal{O}$). The expected degeneracies are then 1, 2 and 3

(4.9)
(1) $Z_7$is Abelian and cyclic. Finding its seven 1-dimensional inequivalent irreducible representations amounts to finding the solutions of the equation$\alpha^7 = 1,\ \alpha \in \mathbf{C}$
(2) Z_6 is Abelian but not cyclic. The group of the equilateral triangle is not Abelian

(4.10)
(1) One has $g_1Hg_2H = g_1g_2\big(g_2^{-1}Hg_2\big)H = g_1g_2H$, indeed $g_2^{-1}h\,g_2 = h' \in H, \ldots$
(2) $GL_n(\mathbf{C}) = SL_n(\mathbf{C}) \times \mathbf{C}$, where $K = SL_n(\mathbf{C})$ is the "special" subgroup of matrices M_1 with $\det M_1 = 1$; indeed, if $\det M \neq 1$, one can write $M = \lambda\, I \cdot M_1 = M_1 \cdot \lambda\, I$ where $\lambda^n = \det M$
(3) $U_n = SU_n \times U_1$
(4) $K = SL_n(\mathbf{R})$ and $GL_n(\mathbf{R})/SL_n(\mathbf{R}) \simeq \mathbf{R}$, but one has $GL_n(\mathbf{R}) = SL_n(\mathbf{R}) \times \mathbf{R}$ *only if* n is odd: indeed, if n is even, there are no real matrices commuting with $SL_n(\mathbf{R})$, and therefore of the form λI, such that $\lambda^n < 0$. Similarly, $O_n = SO_n \times Z_2$, where Z_2 is (isomorphic to) $\{1, -1\}$, *only if* n is odd

(4.11)
(1) The eigenvectors are $(1, \pm i)$
(2) $\exp(im\varphi)$, $m \in \mathbf{Z}$; only $\exp(\pm i\varphi)$ are faithful. Apart from the trivial one, all the irreducible representations of O_2 are 2-dimensional
(3) The circular polarizations

(4.12)
The eigenvectors lie along the light "cones", or better, in the space x, t, the light "lines" $x = \pm ct$

(4.13)
(1) A "constructive" procedure (useful also in more general cases) is the following: diagonalize M by means of a change of basis (unitary in this case): $M = U\, M_{diag} U^{-1}$; the elements in M_{diag} have the form $\exp(i\,\alpha_j)$; $j = 1, \dots, n$; with $\alpha_j \in \mathbf{R}$; then write $M_{diag} = \exp(A_{diag})$ where $A_{diag} = \dots$ and $A^+_{diag} = -A_{diag}$. Show now that, for any matrix A and any invertible matrix S one has

$$S\,(\exp\,A)S^{-1} = \exp(S\,A\,S^{-1})$$

and then conclude $M = \exp A$. Show that $M\,M^+ = M^+ M = I$ and $\det\,M = 1$ give just $A = -A^+$ and $\mathrm{Tr}\,A = 0$. The dimension (over the reals) of a complex $n \times n$ matrix is $2n^2$; the dimensions of SU_n and of U_n are resp. $n^2 - 1$ and n^2
(2) In general, the correspondence between the group G and the Lie algebra $\mathscr{A}$ is continuous and one-to-one in a neighborhood of the identity of the group and the "zero" of the algebra (the origin of $\mathscr{A}$ viewed as linear space). The matrices in O_n have determinant either $+1$ or -1; those with $\det = -1$ cannot be continuously connected with the matrices having $\det = +1$; the former belong to a manifold not connected with the manifold of the latter, which contains the identity. The dimension of SO_n is $n(n-1)/2$

(4.14)
(1) No: take, e.g.., the subgroups generated by the matrices A_1, A_2 of the rotation group, see Problem 4.28, q.(2)

(4.15)

(1) $\quad A = \begin{pmatrix} 0 & -1 \\ 1 & 0 \end{pmatrix} \quad ; \quad$ (2) $\quad A = \begin{pmatrix} 0 & -1 \\ -1 & 0 \end{pmatrix}$

(3) E.g., with $A = \begin{pmatrix} 0 & -1 \\ 1 & 0 \end{pmatrix}$,

$$\exp(a\,A) = \Big(I + a\,A - \frac{a^2}{2!} I - \frac{a^3}{3!} A + \dots\Big) = I\Big(1 - \frac{a^2}{2!} + \frac{a^4}{4!} + \dots\Big) + A\Big(a - \frac{a^3}{3!} + \dots\Big) =$$

$$I\,\cos a + A\,\sin a = \begin{pmatrix} \cos a & -\sin a \\ \sin a & \cos a \end{pmatrix}$$

(4.16)
The first group is a 1-dimensional dilation; the second a rotation, or the group U_1; the next generates dilations in $\mathbf{R}^2$; the two last algebras of the first line can be viewed, for instance, as the generators of a periodic motion (closed orbit) and respectively of a non-periodic motion (dense orbit) on a torus. The first two algebras of the second

line, considered as transformations on $\mathbf{R}^2$ (resp. $\mathbf{R}^3$), describe a diverging spiral in the plane and a spiral on a cylinder. The last case can be interpreted, e.g., as the group of time-evolution $u(t) = \exp(At)u_0$ of the dynamical system $\dot{u} = Au$, where $u = u(t) \in \mathbf{R}^n$, with a given initial condition $u(0) = u_0 \in \mathbf{R}^n$, and where the time t plays the role of Lie parameter

(4.17)
The composition rule is $(\mathscr{O}_2, \mathbf{b}_2) \cdot (\mathscr{O}_1, \mathbf{b}_1) = (\mathscr{O}_2\,\mathscr{O}_1, \mathscr{O}_2\mathbf{b}_1 + \mathbf{b}_2)$.

$$A_3 = \begin{pmatrix} 0 & -1 & 0 & 0 \\ 1 & 0 & 0 & 0 \\ 0 & 0 & 0 & 0 \\ 0 & 0 & 0 & 0 \end{pmatrix}, \text{ etc. ; } B_1 = \begin{pmatrix} 0 & 0 & 0 & 1 \\ 0 & 0 & 0 & 0 \\ 0 & 0 & 0 & 0 \\ 0 & 0 & 0 & 0 \end{pmatrix}, \text{ etc.}$$

(4.18)
(1) *(a)* $A = -d/dx$
(b) $A = \begin{pmatrix} 0 & 1 \\ 0 & 0 \end{pmatrix}$ for case (α); $A = i\lambda$ and $A = \mu$, resp. for cases (γ) and (δ).

Only $i\lambda$ is anti-Hermitian (together with $-d/dx$ of case (β), of course), in agreement with Stone theorem
(2) $A = y(\partial/\partial x) - x(\partial/\partial y) = -d/d\varphi$; (3) $A = t(\partial/\partial x) + x(\partial/\partial t)$

(4.19)
The representation (α) is not unitary, partially reducible, faithful.
(β) is unitary, reducible, faithful.
(γ) is unitary, irreducible, not faithful (indeed, if $b = a + 2n\pi/\lambda, \ldots$).
(δ) is not unitary, irreducible, faithful (if $\mu \neq 0$, of course)

(4.20)
(1) It is easy to verify that the group is not Abelian

(4.21)
(1) $D = -x\, d/dx$; $D^+ = I + x\, d/dx$; (2) $\widetilde{D} = -(x\, d/dx + 1/2) = -\widetilde{D}^+$

(4.22)
(1) In general, $[A_1, A_2] = a_1A_1 + a_2A_2$. If $a_1 = 0$, $a_2 \neq 0$, replace A_1 with $A' = a_2^{-1}A_1$, then $[A', A_2] = \ldots$. If $a_1a_2 \neq 0$, put $A'' = a_1A_1 + a_2A_2$, then $[A_1, A''] = \ldots$

(4.23)
The fourth generator generates the "hyperbolic rotations": see Problems 4.12, 4.18, q.(3), with the variable t replaced by y

(4.24)
(1) The translations generate 2-dimensional Abelian invariant subalgebras
(1–4) Apart from the group (1)(i), all the groups (1–2–3) are subgroups of the group (4). The algebra of this is the following. Let A_i, B_i, P_i, $(i = 1, 2, 3)$, and H denote

resp. the rotations, the Lorentz boosts, the space translations, and the time translation: e.g., using differential notation,

$$A_1 = z\,\partial/\partial y - y\,\partial/\partial z \quad , \quad B_1 = x\,\partial/\partial t + t\,\partial/\partial x \quad , \quad P_1 = \partial/\partial x \quad , \quad \text{etc.}$$

and $H = \partial/\partial t$. Then

$$[\,A_i\,,\,A_j] = \varepsilon_{ijk}A_k \quad , \quad [\,B_i\,,\,B_j] = -\varepsilon_{ijk}A_k \quad , \quad [\,A_i\,,\,B_j] = \varepsilon_{ijk}B_k$$

$$[\,A_i\,,\,P_j] = \varepsilon_{ijk}P_k \quad , \quad [\,B_i\,,\,P_i] = -H \quad , \quad [\,B_i\,,\,H] = -P_i$$

and the other commutators vanish; in particular, P_i and H generate a 4-dimensional Abelian invariant subalgebra

(4.25)
(2) This algebra is isomorphic to that in q.(2) *(ii)* of the previous problem

(4.26)
(2) The generators are $A_1 = -d/dx$, $A_2 = i\,x$, $A_3 = i$
(3) The Fourier transform of the map given in (2) is

$$\exp(-ia_3)\exp(ia_1a_2)\exp(ika_1)\widehat{f}(k+a_2)$$

which is quite similar to the expression of $\big(U_{\mathbf{a}}f\big)(x)$. Indeed, the Heisenberg group can be seen as the group of translations in the space x and (independent) translations in the space k of the Fourier transforms (or in the momentum space, in the physical interpretation), and there are no invariant subspaces

(4.27)
(1) There is a 1-dimensional invariant subspace $r^2 = x^2 + y^2 + z^2$; the other invariant subspace is the 5-dimensional space generated by the 5 spherical harmonics $Y_{\ell,m}(\theta,\varphi)$ with $\ell = 2$: e.g., $z^2 \propto Y_{2,0} \propto \cos^2\theta$; $x^2 - y^2 \propto Y_{2,2} + Y_{2,-2} \propto \sin^2\theta \cdot \cos 2\varphi$
(2) This 10-dimensional space is a superposition of spherical harmonics $Y_{\ell,m}(\theta,\varphi)$ with $\ell = 1$ and $\ell = 3$

(4.28)
(1) $A_i'' = \lambda^{-1}A_i'$
(2) $C = -2$: in quantum mechanics the matrices $i\,A_j$, $j = 1, 2, 3$, describe (apart from the factor $\hbar$) the components of the angular momentum $\ell = 1$ and then $i^2\,C = \ell(\ell+1) = 2$; similarly, $C' = -3$: indeed, the matrices $(i/2)\,A_j'$ correspond to the components of the spin $j = 1/2$ and $(i/2)^2C' = j(j+1) = 3/4$
(3) $a = 0, \pm 2\pi, \ldots$; $a'' = 0, \pm 4\pi, \ldots$

(4.29)
(1) The center of SU_n contains n elements. The center is clearly an invariant subgroup
(2) It is known that for any integer N there is an irreducible representation of SU_2 of dimension N, which is related to an angular momentum or spin $j = 0,\ 1/2,\ 1, \ldots,$ with $N = 2j + 1$; the odd-dimensional representations are also the (faithful) irreducible representations of SO_3 (apart from $N = 1$, of course). Therefore, the odd-dimensional representations of SU_2 are not faithful representations of SU_2. In particular SO_3 has no center

(4.30)
(1) The invariant subspaces are given by the antisymmetric matrices (3-dim.), the traceless symmetric matrices (5-dim.) and the "traces", i.e., the multiples of the identity (1-dim.). In symbols: $(\ell = 1) \otimes (\ell = 1) = (\ell = 0) \oplus (\ell = 1) \oplus (\ell = 2) = \underline{1} \oplus \underline{3} \oplus \underline{5}$

(4.31)
(1) $\underline{3} \otimes \underline{3}$ decomposes into the 6-dimensional representation acting on the symmetric tensors $T_{(ij)}$ and the 3-dimensional representation on the antisymmetric tensors $T_{[ij]}$, equivalent to the vectors $z^k = \varepsilon^{ijk} T_{[ij]}$, or : $\underline{3} \otimes \underline{3} = \underline{6} \oplus \underline{3}^*$.

The tensor product $\underline{3} \otimes \underline{3}^*$ decomposes into the 8-dimensional representation of the traceless tensors T^i_j and the 1-dimensional "traces": $\underline{3} \otimes \underline{3} = \underline{8} \oplus \underline{1}$
(2) $\underline{3} \otimes \underline{3} \otimes \underline{3} = (\underline{3} \otimes \underline{3}) \otimes \underline{3} = (\underline{3}^* \oplus \underline{6}) \otimes \underline{3} = \underline{1} \oplus \underline{8} \oplus \underline{8} \oplus \underline{10}$ where $\underline{10}$ is the 10-dimensional representation acting on the symmetric tensors $T_{(ijk)}$

(4.32)
E.g., $u = \sin x \exp(-y)$ and $u = \cosh(ax + by) \cos(bx - ay)$, $\forall a, b$ are other solutions

(4.33)
$f(r) = c_1 \log r + r^{n+2}/(n+2)^2 + c_2$, where c_i are arbitrary constants

(4.34)
(1) The ODE is $f_{ss} + v f_s = 0$, with solution $u(x,t) = c_1 \exp\big(-v(x - vt)\big) + c_2$, for any v
(2) Apart from the trivial solution $u = x - vt$ with arbitrary v, the only admitted velocities are $v = \pm 1$, with the well-known solutions $u = f_1(x - t) + f_2(x + t)$, with arbitrary f_1, f_2

(4.35)
(2) The ODE is $4s f_{ss} + 4 f_s = s^n$, with solution

$$u(x,t) = c_1 \log|x^2 - t^2| + \frac{(x^2 - t^2)^{n+1}}{4(n+1)^2} + c_2$$

(4.36)

The irreducible representations of SO_3 have odd dimension $2\ell+1$ ($\ell=0,1,2,\ldots$) as well-known, then the energy levels of systems exhibiting spherical symmetry are expected to have *odd* degeneracy. In the presence of a magnetic field, the surviving symmetry is SO_2 (the rotations around the magnetic field, no reflections), which admits only 1-dimensional irreducible representations. Then, Schur lemma indicates that the degeneracy of the levels should be completely removed (Zeeman effect).

In the presence of an electric field, the surviving symmetry is O_2, which admits 1 and 2-dimensional irreducible representations. Then, ...(Stark effect)

(4.37)

(3) The irreducible representations are described by the couple $j_M,\ j_N$ with dimension $(2j_M+1)(2j_N+1)$

(4) A_3 and B_3 are two commuting operators, and no other operators commute with them. The Casimir operators: $C_1 = 2j_M(j_M+1) + 2j_M(j_M+1);\ C_2 = \ldots$

(4.38)

(1) The representations involved are those with $j_M = j_N$. The degeneracies are then $(2j_M+1)^2 = n^2,\ n = 1,2,\ldots$

(2) $E_n = -me^4/2\hbar^2 n^2,\ n = 1,2,\ldots$, as well-known

(3) The last result can be shown, e.g.., by induction (recall $\ell_{Max} = n-1$):

$$\sum^{n} \cdots = \sum^{n-1} \cdots + 2n+1 = n^2 + (2n+1) = (n+1)^2$$

(4.39)

(2)(*b*) The commutation rules of the algebra of SU_3 can be obtained from its faithful representation by means of 3×3 matrices as indicated in Problem 4.13, q.(1).

E.g., $A_1^2 - A_2^1 = i(x_2\partial/\partial x_1 - x_1\partial/\partial x_2)$

(2)(*c*) $\eta_1 v_0(x_1) = x_1 v_0 + \partial v_0/\partial x_1 = 0$, then $v_0(x_1) \propto \exp(-x_1^2/2)$.

Next: $[A_1^1, \eta_1^+] = \eta_1^+[\eta_1, \eta_1^+] = \eta_1^+$, and $v_1(x_1) \equiv \eta_1^+ v_0(x_1) \propto x_1 v_0(x_1)$ satisfies $A_1^1 v_1 = v_1$; $v_2(x_1) \equiv \eta_1^+ v_1(x_1)$ satisfies $A_1^1 v_2 = 2v_2$, etc.

The eigenvalues of $\mathtt{Tr}\,\mathbf{A}$ are 0,1,2,.... The states with $E = 5/2$ are given by $\eta_1^+ u_0 \propto x_1 \exp(-r^2/2)$, $\eta_2^+ u_0 \propto x_2 \exp(-r^2/2)$, $\eta_3^+ u_0 \propto \ldots$

(2)(*e*) The 6-dimensional representation contains $\ell = 0$ (1-dimensional) and $\ell = 2$ (5-dimensional)

Bibliography

Books covering several topics in Mathematical Methods of Physics

1. G.B. Arfken, H.J. Weber, F.E. Harris, *Mathematical Methods for Physicists* (Academic Press, San Diego, 2013)
2. F.W. Byron, R.W. Fuller, *Mathematics of Classical and Quantum Mechanics* (Dover, New York, 1969)
3. R. Courant, D. Hilbert, *Methods of Mathematical Physics* (Wiley, New York, 1989)
4. S. Hassani, *Mathematical Physics* (Springer, New York, 1999)
5. P.M. Morse, H. Feshbach, *Methods of Theoretical Physics* (McGraw-Hill, New York, 1953)
6. M. Reed, B. Simon, *Methods of Modern Mathematical Physics* (Academic, New York, 1972)
7. K.F. Riley, M.P. Hobson, S.J. Bence, *Mathematical Methods for Physics and Engineering: A Comprehensive Guide* (Cambridge University Press, Cambridge, 2006)
8. L. Schwartz, *Mathematics for the physical sciences* (Hermann, Paris, 1966)
9. W.-H. Steeb, *Problems and solutions in Theoretical and Mathematical Physics* (Word Scientific, Singapore, 2009)
10. M.E. Taylor, *Partial Differential Equations, Basic Theory* (Springer, New York, 1996)
11. H.F. Weinberger, *A First Course on Partial Differential Equations* (Blaisdell, Waltham, 1965)

Books covering particular arguments

12. N. Akhiezer, I.M. Glazman, *Theory of Linear Operators in Hilbert Spaces* (Ungar, New York, 1963)
13. F. Riesz, B. Nagy: *Functional Analysis* (Dover New York, 1990)
14. H.L. Royden, *Real Analysis* (Macmillan Co., New York, 1963)
15. W. Rudin, *Principles of Mathematical Analysis* (McGraw-Hill, New York, 1984)
16. K. Yosida, *Functional Analysis* (Springer, New York, 1965)
17. L.V. Ahlfors, *Complex Analysis* (Mc Graw-Hill, New York, 1966)

G. Cicogna, *Exercises and Problems in Mathematical Methods of Physics*, Undergraduate Lecture Notes in Physics,
https://doi.org/10.1007/978-3-030-59472-5

18. M. Lavrentiev. B. Chabat *Methods of the theory of function of complex variable* (Nauka. Moscow, 1987)
19. A.I. Markushevitch, *Theory of a Complex Variable* (Mc Graw-Hill, New York, 1965)
20. W. Rudin, *Real and Complex Analysis* (McGraw-Hill, New York, 1987)
21. A. Papoulis, *The Fourier Integral and its Applications* (McGraw-Hill, New York, 1962)
22. A. Vretblad, *Fourier Analysis and its Applications* (Springer, Berlin, 2003)
23. V. Vladimirov, *Generalized Functions in Mathematical Physics* (Mir Publishers, Moscow, 1979)
24. H. Georgi, *Lie Algebras in Particle Physics* (Benjamin Cummings Publ. Co., London, 1982)
25. B.C. Hall, *Lie Groups, Lie Algebras and Representations* (Springer, Berlin, 2003)
26. M. Hamermesh, *Group Theory and Applications to Physical Problems* (Addison Wesley, London, 1962)
27. P.J. Olver, *Application of Lie Groups to Differential Equations* (Springer, New York, 1993)
28. D.H. Sattinger, O.L. Weaver, *Lie groups and Algebras with Applications to Physics, Geometry and Mechanics* (Springer, New York, 1986)
29. S. Sternberg, *Group Theory and Physics* (Cambridge University Press, Cambridge, 1994)

Printed in the United States
By Bookmasters